GUIDEBOOK TO COMMERCIAL POLYMERS

POLYMER PROCESS ENGINEERING SERIES

by Nicholas P. Cheremisinoff

Guidebook to Commercial Polymers: Properties and Applications
Guidebook to Extrusion Technology
Guidebook to Mixing and Compounding Practices
Guidebook to Film Blowing and Coating Operations
Guidebook to Plastics Recycling and Disposal Technology

GUIDEBOOK TO COMMERCIAL POLYMERS

Properties and Applications

Nicholas P. Cheremisinoff
SciTech Technical Services

P T R Prentice Hall
Englewood Cliffs, New Jersey 07632

Library of Congress Cataloging-in-Publication Data

Cheremisinoff, Nicholas P.
Guidebook to commercial polymers: properties and applications / Nicholas P. Cheremisinoff.
p. cm. -- (Polymer process engineering series)
Includes index.
ISBN 0-13-175613-3
1. Polymers. I. Title. II. Series
TP1087.C48 1993 93-13060
620.1'92--dc20 CIP

Editorial/production supervision: *bookworks*
Cover design: *Jeannette Jacobs*
Acquisitions editor: *Betty Sun*
Buyer: *Mary Elizabeth McCartney*

© 1993 by P T R Prentice-Hall, Inc.
A Simon & Schuster Company
Englewood Cliffs, New Jersey 07632

The publisher offers discounts on this book when ordered in bulk quantities. For more information, contact:

Corporate Sales Department
PTR Prentice Hall
113 Sylvan Avenue
Englewood Cliffs, NJ 07632

Phone: 201-592-2863
Fax: 201-592-2249

All rights reserved. No part of this book may be reproduced, in any form or by any means, without permission in writing from the publisher.

Printed in the United States of America
10 9 8 7 6 5 4 3 2 1

ISBN 0-13-175613-3

Prentice-Hall International (UK) Limited, *London*
Prentice-Hall of Australia Pty. Limited, *Sydney*
Prentice-Hall Canada Inc., *Toronto*
Prentice-Hall Hispanoamericana, S.A., *Mexico*
Prentice-Hall of India Private Limited, *New Delhi*
Prentice-Hall of Japan, Inc., *Tokyo*
Simon & Schuster Asia Pte. Ltd., *Singapore*
Editora Prentice-Hall do Brasil, Ltda., *Rio de Janeiro*

CONTENTS

PREFACE

This volume is written as a desk reference for product development managers, polymer compounders and parts fabricators, technicians, and polymer chemists involved in the selection, design, and use of elastomeric and plastic materials. Emphasis is given to polyolefin-type polymers because of their wide use in today's marketplace, and the high level of product and applications development activities which many polymer suppliers are now emphasizing.

The desk reference is organized into six chapters. Chapter 1 provides an overview of polymer science principles and a review of polymerization chemistry. Chapter 2 covers addition-type polymers (that is, polyolefin) and chemistry processes specific to this important class. Chapter 3 contains general properties data and specific information on high-volume polyolefins such as polypropylene and polyethylene. Specialty elastomers such as ethylene-propylene copolymers and tarpolymers are discussed in Chapter 4. Chapter 5 contains data and information on blends and alloys. Emphasis in this chapter is given to end-user applications and commercial market trends. Chapter 6 contains an extensive compilation of notes on applications of elastomers and plastic blends and alloys in the automotive market segment. This particular industry is highlighted because of the diversity of polymer applications, the high-volume use of elastomers/plastics, and the anticipated growth in market applications.

This desk reference provides an important guide for polymer development specialists who often are missing information on applications and marketing trends.

Nicholas P. Cheremisinoff, Ph.D.

GUIDEBOOK TO COMMERCIAL POLYMERS

1 *PRINCIPLES OF POLYMER SCIENCE*

1.1 BACKGROUND AND BASIC DEFINITIONS

Polymeric materials can be defined simply as chain-like molecules that are formed by linking together many small molecules (monomers) through a repetitive series of chemical reactions. Examples of naturally occurring polymers are:

proteins (polyaminoacids)
polynucleotides (DNA, RNA)
polysaccharides (cellulose)
natural rubber (heavea)

Examples of synthetically produced polymers are:

nylon (polyamides)
polyethylene
polystyrene
polybutadiene
epoxy resins

The term *polymerization* refers to a branch of organic chemistry making use of standard organic reactions to form macromolecules.

A simple example of polymerization is the synthesis of nylon. Amide links are formed by the reaction of carboxylic acid groups and amine groups:

$$\underset{\text{acid}}{-\overset{\overset{\displaystyle O}{\|}}{C}-OH} + \underset{\text{primary amine}}{-NH_2} \xrightarrow{\Delta} H_2O\uparrow + \underset{\text{amide}}{-\overset{\overset{\displaystyle O}{\|}}{C}-\overset{\overset{\displaystyle H}{|}}{N}-}$$

This results in the formation of the polymer Nylon 6,6:

$$HO-\overset{O}{\overset{\|}{C}}-(CH_2)_4-\overset{O}{\overset{\|}{C}}-OH \qquad H_2N-(CH_2)_6-NH_2$$

adipic acid hexamethylene diamine

$$\Downarrow \Delta,\ -H_2O$$

$$\text{etc}-\overset{O}{\overset{\|}{C}}-(CH_2)_4-\overset{O}{\overset{\|}{C}}-\overset{H}{\overset{|}{N}}-(CH_2)_6-\overset{H}{\overset{|}{N}}-\overset{O}{\overset{\|}{C}}-(CH_2)_4-\overset{O}{\overset{\|}{C}}-\overset{H}{\overset{|}{N}}-(CH_2)_6-\overset{H}{\overset{|}{N}}-\text{etc}$$

As the chemical reaction proceeds, molecular chains get longer. Each molecule has unreacted functional groups (acid or amine) at the chain ends. Growth stops when all the acid groups or all the amine groups are reacted, or if cooled before a complete reaction occurs.

An important type of polymerization is condensation or step growth. An example of this type is the synthesis of polystyrene. In polystyrene synthesis, C-C (carbon-carbon) bonds formed by the reaction of free radicals and vinyl groups take place:

$$R\cdot + \;(H)(H)C{=}C(H)(R') \dashrightarrow R-\underset{H}{\overset{H}{C}}-\underset{R'}{\overset{H}{C}}\cdot$$

radical vinyl molecule radical

$$(H)(H)C{=}C(H)(C_6H_5) \xrightarrow[\text{peroxide (trace)}]{\Delta} \text{etc}-\underset{H}{\overset{H}{C}}-\underset{H}{\overset{C_6H_5}{C}}-\underset{H}{\overset{H}{C}}-\underset{H}{\overset{C_6H_5}{C}}-\text{etc}$$

Chains can also be formed by the rapid addition of monomer molecules to free radicals formed by the slow decomposition of peroxide molecules. A radical continues to add monomer (as previously) until it encounters another radical and reacts to give an inactive group (for example, by coupling). This is known as addition or chain-growth polymerization. An example of this type of polymerization is as follows:

step growth:	all chains grow simultaneously
chain growth:	a few chains grow at a time until all monomer (or peroxide) is consumed

Products in both cases are stable macromolecules.

We may now introduce the term *monomer*. Any molecule that is capable of forming at least two bonds with other molecules by a chemical reaction is known as a monomer, and has associated with it a degree of functionality, f. For example,

$$CH_3{-}\overset{\overset{\large O}{\|}}{C}{-}OH \qquad \text{monofunctional } (f = 1)$$

$$HO{-}\overset{\overset{\large O}{\|}}{C}{-}(CH_2)_4{-}\overset{\overset{\large O}{\|}}{C}{-}OH \qquad \text{difunctional } (f = 2)$$

$$C\left[CH_2{-}\overset{\overset{\large O}{\|}}{C}{-}OH\right]_4 \qquad \text{tetrafunctional } (f = 4)$$

With respect to amide or ester linking:

$$-NH_2 + -\overset{\overset{\large O}{\|}}{C}{-}OH \xrightarrow[-H_2O]{\Delta} -\overset{\overset{\large H}{|}}{N}{-}\overset{\overset{\large O}{\|}}{C}- \qquad \text{(amide)}$$

or

$$-OH + -\overset{\overset{\large O}{\|}}{C}{-}OH \xrightarrow[-H_2O]{\Delta} -O{-}\overset{\overset{\large O}{\|}}{C}- \qquad \text{(ester)}$$

Hence, a structure which is monofunctional has a chain end only. A difunctional structure has linear chains of any length. Multifunctional ($f > 3$) refers to structures with branched chains and networks, such as:

$$(H)(H)C{=}C(H)(R) \qquad f = 2$$

vinyl compound

$$(H)(H)C{=}C(H){-}(CH_2)_3{-}C(H){=}C(H)(H) \qquad f = 4$$

divinyl compound

Reactions that open up one bond of a double bond can form two new single bonds. Vinyl monomers give linear chains; divinyl monomers can give branched chains. For example,

$$(H)(H)C{=}C{-}\overset{\overset{\large O}{\|}}{C}{-}OH$$

$f = 1$ for acid group $r \times n$.

$f = 2$ for vinyl group $r \times n$.

acrylic acid

Mers are the molecular units of the chain derived from the individual monomers. These are sometimes the same as the repeating unit. As examples:

$$-\mathrm{CH_2}-\mathrm{CH}(\mathrm{C_6H_5})- $$

is the polystyrene *mer* and *repeat unit,*

$$[-\overset{\mathrm{O}}{\overset{\|}{\mathrm{C}}}-(\mathrm{CH_2})_4-\overset{\mathrm{O}}{\overset{\|}{\mathrm{C}}}-\overset{\mathrm{H}}{\overset{|}{\mathrm{N}}}-(\mathrm{CH})_6-\overset{\mathrm{H}}{\overset{|}{\mathrm{N}}}-]$$

is the Nylon 6,6 repeat unit.

and

$$\left[-\overset{\mathrm{O}}{\overset{\|}{\mathrm{C}}}-(\mathrm{CH_2})_4-\overset{\mathrm{O}}{\overset{\|}{\mathrm{C}}}-\right] \text{ and } \left[-\overset{\mathrm{H}}{\overset{|}{\mathrm{N}}}-(\mathrm{CH_2})_6-\overset{\mathrm{H}}{\overset{|}{\mathrm{N}}}-\right]$$

are the mers.

Examples of common polymers and their structures are given in Table 1-1.

Since polymers are simply very large, chain-like organic molecules, their units obey the same structural laws and have similar bonding energy, intermolecular attractions, and chemical reactivity as their small-molecule analogs. The term *chemical microstructure* is used to distinguish between local structure of the chains (on the mer scale) from macromolecular structure (large-scale features such as polymer chain length, molecular weight, molecular weight distribution, and long chain branching). *Physical microstructure* refers to how the chains pack together (that is, whether ordered to form crystalline segments or disordered to form liquid-like or amorphous regions). *Morphological structure* describes features even larger than macromolecular dimensions (size, shape, and arrangement of the crystalline regions in crystallized polymers; size and shape of the dispersed phases in polymer blends).

The physical properties of polymers and, hence, their macroscopic behavior depend on all the preceding structural features. Furthermore, their effects are not independent of one another. For example, the physical microstructure depends on the chemical microstructure, and morphology can be related to the macromolecular structure as well as the chemical microstructure and the thermal and mechanical history of the material. Structure-property relationships are therefore critical elements in establishing the end-user performance properties of any polymer.

The chemical bonds in organic molecules are for the most part covalent and built around the remarkable capacity of carbon atoms to bond to other carbon atoms. The term *covalent bonding* means that there is more or less equal sharing of an electron pair by two atoms. (Ionic bonding involves electron pairing also, but the pair resides mostly on only one of the atoms.)

A carbon atom has four electrons in its outer shell and can form up to four covalent bonds with other atoms to reach a completed outer shell of eight

Table 1-1
Examples of Common Polymers and Their Structures

	Monomer	Polymer	Usual Method of Polymerization	Typical Degree of Crystalline Order
1.	$H_2C{=}CHCl$ vinyl chloride	$-[CH_2-CHCl]-$ polyvinyl chloride	free radical	microcrystalline
2.	$H_2C{=}CH_2$ ethylene	$-[CH_2-CH_2]-$ polyethylene	free radical and Ziegler–Natta	crystalline
3.	$H_2C{=}CH(CH_3)$ propylene	$-[CH_2-CH(CH_3)]-$ polypropylene	Ziegler–Natta	crystalline
4.	$H_2C{=}CH(C_6H_5)$ styrene	$-[CH_2-CH(C_6H_5)]-$ polystyrene	free radical	amorphous
5.	$H_2C{=}CH-O-C({=}O)-CH_3$ vinyl acetate	$-[CH_2-CH(O-C({=}O)-CH_3)]-$ polyvinyl acetate	free radical	amorphous

Table 1-1
Examples of Common Polymers and Their Structures *(Continued)*

	Monomer	Polymer	Usual Method of Polymerization	Typical Degree of Crystalline Order
6.	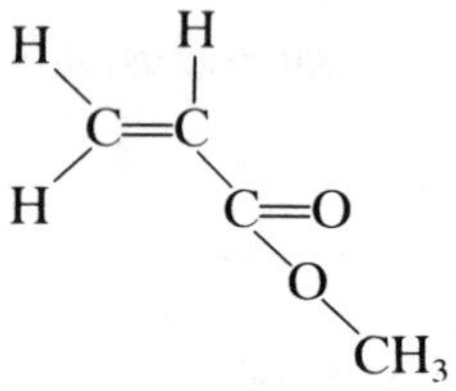 methyl acrylate	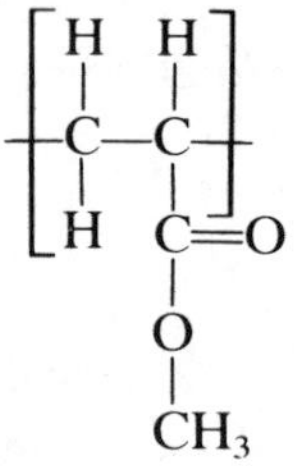 polymethyl acrylate	free radical	amorphous
7.	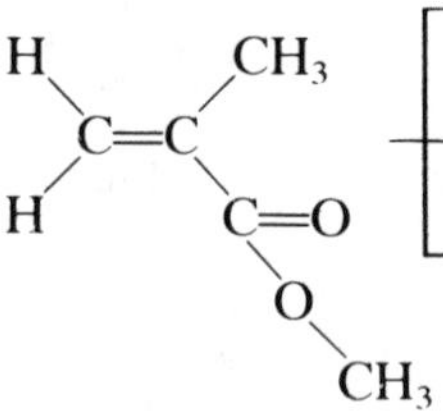 methyl methacrylate	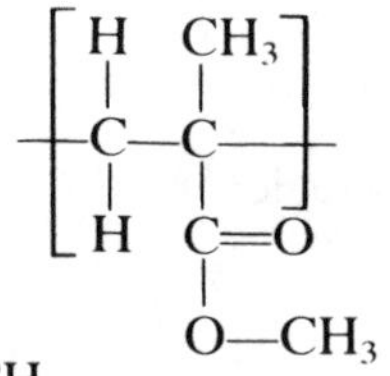 polymethyl methacrylate	free radical	amorphous
8.	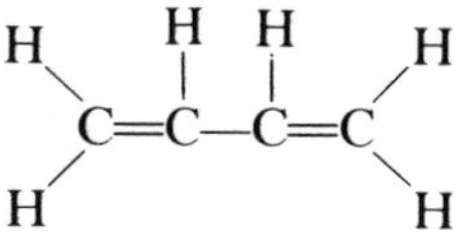 butadiene	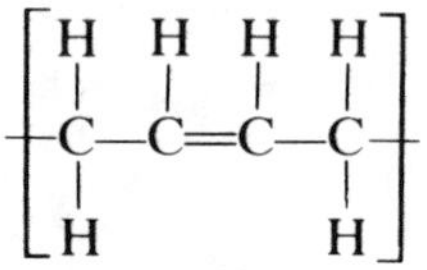 polybutadiene	free radical or anionic	amorphous
9.	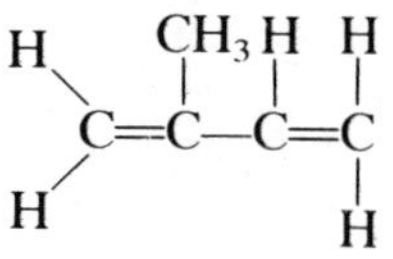 isoprene	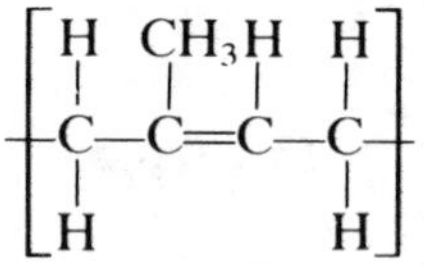 polyisoprene	free radical or anionic	amorphous
10.	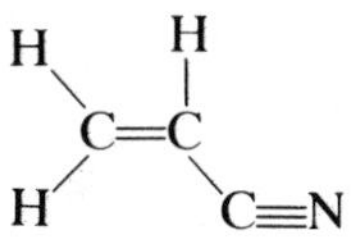 acrylonitrile	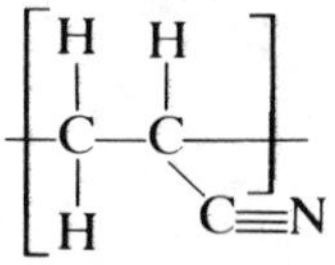 polyacrylonitrile	free radical	crystalline

Table 1-1
Examples of Common Polymers and Their Structures *(Continued)*

	Monomer	Polymer	Usual Method of Polymerization	Typical Degree of Crystalline Order
11.	$\left[\begin{matrix} H & & H \\ & C{=}C & \\ H & & OH \end{matrix} \right]$ vinyl alcohol (hypothetical)	$\left[-CH_2-CH(OH)- \right]$ polyvinyl alcohol	hydrolysis of polyvinyl acetate	crystalline
12.	$H_2C{=}O$ formaldehyde	$\left[-CH_2-O- \right]$ polyformaldehyde (Delrin or Celcon)	cationic	crystalline
13.	$H_2N-(CH_2)_6-NH_2$ hexamethylene diamine; $HO-\overset{O}{\overset{\parallel}{C}}-(CH_2)_4-\overset{O}{\overset{\parallel}{C}}-OH$ adipic acid	$\left[-\overset{O}{\overset{\parallel}{C}}-(CH_2)_4-\overset{O}{\overset{\parallel}{C}}-N(H)-(CH_2)_6-N(H)- \right]$ polyamides (Nylon 6,6)	condensation	crystalline
14.	phenol (C_6H_5OH); formaldehyde ($H_2C{=}O$)	$\left[-CH_2-C_6H_2(OH)(-CH_2-)-CH_2- \right]$ phenolic resins	condensation	amorphous

electrons. Hydrogen has a single electron and can form one covalent bond to reach its completed shell of two electrons. An element's location in the periodic table governs the bonding capacity of that element. Thus,

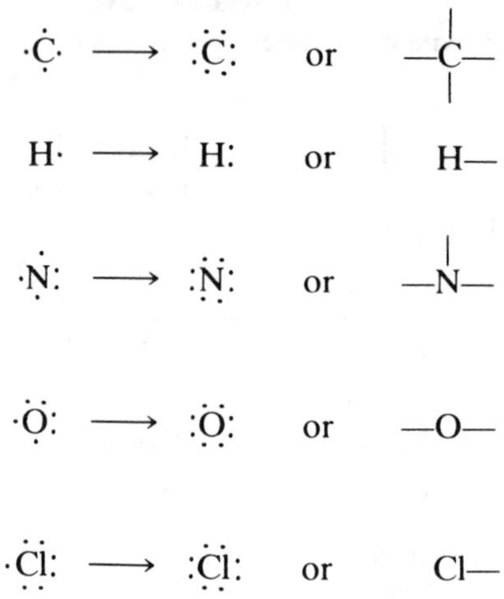

·Ċ· ⟶ :C̈: or —C— (with bonds above and below)

H· ⟶ H: or H—

·N̈: ⟶ :N̈: or —N— (with bond above)

·Ö: ⟶ :Ö: or —O—

·C̈l: ⟶ :C̈l: or Cl—

Atomic weights of elements are given in Table 1-2, and the periodic table is given in Table 1-3.

The number of covalent bonds is referred to as the *valence*. The elements C, H, N, O, Cl, and S have valences of 4, 1, 3, 2, 1, and 2, respectively, in most organic molecules. For many atoms, multiple bonds like the following ones are common.

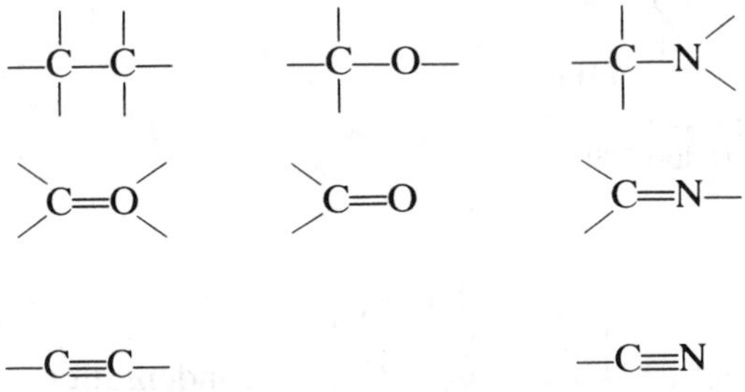

—C—C—   —C—O—   —C—N<

>C=C<   >C=O   >C=N—

—C≡C—   —C≡N

It's important to note that the sum of bonds for any atom must be equal to the valence of the element. Species such as free radicals (unpaired electrons, —C· or :C·) are inherently unstable and react rapidly to form normal electron- paired bonds, for example,

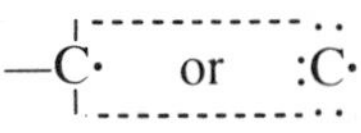

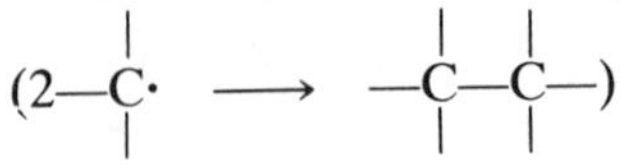

(2—C· ⟶ —C—C—)

Table 1-2 Atomic Weights

Name	Symbol	At. No.	International atomic weight		Valence
			1961	1959	
Actinium	Ac	89	....	(227)	....
Aluminum	Al	13	26.9815	26.98	3
Americium	Am	95	....	(243)	3, 4, 5, 6
Antimony, stibium	Sb	51	121.75	121.76	3, 5
Argon	Ar	18	39.948	39.944	0
Arsenic	As	33	74.9216	74.92	3, 5
Astatine	At	85	....	(210)	1, 3, 5, 7
Barium	Ba	56	137.34	137.36	2
Berkelium	Bk	97	....	(249)	3, 4
Beryllium	Be	4	9.0122	9.013	2
Bismuth	Bi	83	208.980	208.99	3, 5
Boron	B	5	10.811	10.82	3
Bromine	Br	35	79.909	79.916	1, 3, 5, 7
Cadmium	Cd	48	112.40	112.41	2
Calcium	Ca	20	40.08	40.08	2
Californium	Cf	98	....	(251)	....
Carbon	C	6	12.01115	12.011	2, 4
Cerium	Ce	58	140.12	140.13	3, 4
Cesium	Cs	55	132.905	132.91	1
Chlorine	Cl	17	35.453	35.457	1, 3, 5, 7
Chromium	Cr	24	51.996	52.01	2, 3, 6
Cobalt	Co	27	58.9332	58.94	2, 3
Columbium, see *Niobium*					
Copper	Cu	29	63.54	63.54	1, 2
Curium	Cm	96	....	(247)	3
Dysprosium	Dy	66	162.50	162.51	3
Einsteinium	Es	99	....	(254)	....
Erbium	Er	68	167.26	167.27	3
Europium	Eu	63	151.96	152.0	2, 3
Fermium	Fm	100	....	(253)	....
Fluorine	F	9	18.9984	19.00	1
Francium	Fr	87	....	(223)	1
Gadolinium	Gd	64	157.25	157.26	3
Gallium	Ga	31	69.72	69.72	2, 3
Germanium	Ge	32	72.59	72.60	4
Gold, aurum	Au	79	196.967	197.0	1, 3
Hafnium	Hf	72	178.49	178.50	4
Helium	He	2	4.0026	4.003	0
Holmium	Ho	67	164.930	164.94	3
Hydrogen	H	1	1.00797	1.0080	1
Indium	In	49	114.82	114.82	3
Iodine	I	53	126.9044	126.91	1, 3, 5, 7
Iridium	Ir	77	192.2	192.2	3, 4
Iron, ferrum	Fe	26	55.847	55.85	2, 3
Krypton	Kr	36	83.80	83.80	0
Lanthanum	La	57	138.91	138.92	3
Lead, plumbum	Pb	82	207.19	207.21	2, 4
Lithium	Li	3	6.939	6.940	1
Lutetium	Lu	71	174.97	174.99	3
Magnesium	Mg	12	24.312	24.32	2
Manganese	Mn	25	54.9380	54.94	2, 3, 4, 6, 7
Mendelevium	Md	101	....	(256)	....
Mercury, hydrargyrum	Hg	80	200.59	200.61	1, 2
Molybdenum	Mo	42	95.94	95.95	3, 4, 6
Neodymium	Nd	60	144.24	144.27	3

For the sake of completeness all known elements are included in the list. Several of those more recently discovered are represented only by the unstable isotopes. The value in parenthesis in the atomic weight column is, in each case, the mass number of the most stable isotope.**

Table 1-2 Atomic Weights *(Continued)*

Name	Symbol	At. No.	International atomic weight		Valence
			1961	1959	
Neon	Ne	10	20.183	20.183	0
Neptunium	Np	93	....	(237)	4, 5, 6
Nickel	Ni	28	58.71	58.71	2, 3
Niobium (columbium)	Nb	41	92.906	92.91	3, 5
Nitrogen	N	7	14.0067	14.008	3, 5
Nobelium	No	102	....	(254)	....
Osmium	Os	76	190.2	190.2	2, 3, 4, 8
Oxygen	O	8	15.9994	16.000	2
Palladium	Pd	46	106.4	106.4	2, 4, 6
Phosphorus	P	15	30.9738	30.975	3, 5
Platinum	Pt	78	195.09	195.09	2, 4
Plutonium	Pu	94	....	(242)	3, 4, 5, 6
Polonium	Po	84	....	(210)	....
Potassium, kalium	K	19	39.102	39.100	1
Praseodymium	Pr	59	140.907	140.92	3
Promethium	Pm	61	....	(147)	3
Protactinium	Pa	91	....	(231)	....
Radium	Ra	88	....	(226)	2
Radon	Rn	86	....	(222)	0
Rhenium	Re	75	186.2	186.22	....
Rhodium	Rh	45	102.905	102.91	3
Rubidium	Rb	37	85.47	85.48	1
Ruthenium	Ru	44	101.07	101.1	3, 4, 6, 8
Samarium	Sm	62	150.35	150.35	2, 3
Scandium	Sc	21	44.956	44.96	3
Selenium	Se	34	78.96	78.96	2, 4, 6
Silicon	Si	14	28.086	28.09	4
Silver, argentum	Ag	47	107.870	107.873	1
Sodium, natrium	Na	11	22.9898	22.991	1
Strontium	Sr	38	87.62	87.63	2
Sulfur	S	16	32.064	32.066*	2, 4, 6
Tantalum	Ta	73	180.948	180.95	5
Technetium	Tc	43	....	(99)	6, 7
Tellurium	Te	52	127.60	127.61	2, 4, 6
Terbium	Tb	65	158.924	158.93	3
Thallium	Tl	81	204.37	204.39	1, 3
Thorium	Th	90	232.038	(232)	4
Thulium	Tm	69	168.934	168.94	3
Tin, stannum	Sn	50	118.69	118.70	2, 4
Titanium	Ti	22	47.90	47.90	3, 4
Tungsten (wolfram)	W	74	183.85	183.86	6
Uranium	U	92	238.03	238.07	4, 6
Vanadium	V	23	50.942	50.95	3, 5
Xenon	Xe	54	131.30	131.30	0
Ytterbium	Yb	70	173.04	173.04	2, 3
Yttrium	Y	39	88.905	88.91	3
Zinc	Zn	30	65.37	65.38	2
Zirconium	Zr	40	91.22	91.22	4

* Because of natural variations in the relative abundances of the isotopes of sulfur the atomic weight of this element has a range of ±0.003.

** The 1959 atomic weights are based on O_2 = 16.000 whereas those of 1961 are based on the isotope C^{12}.

Table 1-3 Periodic Table of the Elements

KEY TO CHART

Atomic Number →	50 +2 +4 ← Oxidation States
Symbol →	Sn
Atomic Weight →	118.69
	-18-18-4 ← Electron Configuration

Transition Elements — Group 8 — Transition Elements

1a	2a	3b	4b	5b	6b	7b	8			1b	2b
1 +1 −1 H 1.00797 1											
3 +1 Li 6.939 2-1	4 +2 Be 9.0122 2-2										
11 +1 Na 22.9898 2-8-1	12 +2 Mg 24.312 2-8-2										
19 +1 K 39.102 -8-8-1	20 +2 Ca 40.08 -8-8-2	21 +3 Sc 44.956 -8-9-2	22 +2 +3 +4 Ti 47.90 -8-10-2	23 +2 +3 +4 +5 V 50.942 -8-11-2	24 +2 +3 +6 Cr 51.996 -8-13-1	25 +2 +3 +4 +7 Mn 54.9380 -8-13-2	26 +2 +3 Fe 55.847 -8-14-2	27 +2 +3 Co 58.9332 -8-15-2	28 +2 +3 Ni 58.71 -8-16-2	29 +1 +2 Cu 63.54 -8-18-1	30 +2 Zn 65.37 -8-18-2
37 +1 Rb 85.47 -18-8-1	38 +2 Sr 87.62 -18-8-2	39 +3 Y 88.905 -18-9-2	40 +4 Zr 91.22 -18-10-2	41 +3 +5 Nb 92.906 -18-12-1	42 +6 Mo 95.94 -18-13-1	43 +4 +6 +7 Tc (99) -18-13-2	44 +3 Ru 101.07 -18-15-1	45 +3 Rh 102.905 -18-16-1	46 +2 +4 Pd 106.4 -18-18-0	47 +1 Ag 107.870 -18-18-1	48 +2 Cd 112.40 -18-18-2
55 +1 Cs 132.905 -18-8-1	56 +2 Ba 137.34 -18-8-2	57* +3 La 138.91 -18-9-2	72 +4 Hf 178.49 -32-10-2	73 +5 Ta 180.948 -32-11-2	74 +6 W 183.85 -32-12-2	75 +4 +6 +7 Re 186.2 -32-13-2	76 +3 +4 Os 190.2 -32-14-2	77 +3 +4 Ir 192.2 -32-15-2	78 +2 +4 Pt 195.09 -32-16-2	79 +1 +3 Au 196.967 -32-18-1	80 +1 +2 Hg 200.59 -32-18-2
87 +1 Fr (223) -18-8-1	88 +2 Ra (227) -18-8-2	89** +3 Ac (227) -18-9-2									

*Lanthanides	58 +3 +4 Ce 140.12 -19-9-2	59 +3 Pr 140.907 -20-9-2	60 +3 Nd 144.24 -22-8-2	61 +3 Pm (145) -23-8-2	62 +2 +3 Sm 150.35 -24-8-2	63 +2 +3 Eu 151.96 -25-8-2	64 +3 Gd 157.25 -25-9-2	65 +3 Tb 158.924 -26-9-2	66 +3 Dy 162.50 -28-8-2	67 +3 Ho 164.930 -29-8-2
**Actinides	90 +4 Th 232.038 -19-9-2	91 +5 +4 Pa (231) -20-9-2	92 +3 +4 +5 +6 U 238.03 -21-9-2	93 +3 +4 +5 +6 Np (237) -22-9-2	94 +3 +4 +5 +6 Pu (242) -23-9-2	95 +3 +4 +5 +6 Am (243) -24-9-2	96 +3 Cm (245) -25-9-2	97 +3 +4 Bk (249) -26-9-2	98 +3 Cf (249) -28-8-2	99 Es (254) -29-8-2

Table 1-3 Periodic Table of the Elements *(Continued)*

3a	4a	5a	6a	7a	0	Orbit
					2 0 He 4.0026 2	K
5 +3 B 10.811 2-3	6 +2 +4 −4 C 12.01115 2-4	7 +1 +2 +3 +4 +5 −1 −2 −3 N 14.0067 2-5	8 −2 O 15.9994 2-6	9 −1 F 18.9984 2-7	10 0 Ne 20.183 2-8	K-L
13 +3 Al 26.9815 2-8-3	14 +2 +4 −4 Si 28.086 2-8-4	15 +3 +5 −3 P 30.9738 2-8-5	16 +4 +6 −2 S 32.064 2-8-6	17 +1 +5 +7 −1 Cl 35.453 2-8-7	18 0 Ar 39.948 2-8-8	K-L-M
31 +3 Ga 69.72 -8-18-3	32 +2 +4 Ge 72.59 -8-18-4	33 +3 +5 −3 As 74.9216 -8-18-5	34 +4 +6 −2 Se 78.96 -8-18-6	35 +1 +5 −1 Br 79.909 -8-18-7	36 0 Kr 83.80 -8-18-8	-L-M-N
49 +3 In 114.82 -18-18-3	50 +2 +4 Sn 118.69 -18-18-4	51 +3 +5 −3 Sb 121.75 -18-18-5	52 +4 +6 −2 Te 127.60 -18-18-6	53 +1 +5 +7 −1 I 126.9044 -18-18-7	54 0 Xe 131.30 -18-18-8	-M-N-O
81 +1 +3 Tl 204.37 -32-18-3	82 +2 +4 Pb 207.19 -32-18-4	83 +3 +5 Bi 208.980 -32-18-5	84 +2 +4 Po (210) -32-18-6	85 At (210) -32-18-7	86 0 Rn (222) -32-18-8	-N-O-P
						-O-P-Q

				Orbit
68 +3 Er 167.26 -30-8-2	69 +3 Tm 168.934 -31-8-2	70 +2 +3 Yb 173.04 -32-8-2	71 +3 Lu 174.97 -32-9-2	-N-O-P
100 Fm (252) -30-8-2	101 Md (256) -31-8-2	102 (254) -32-8-2	103 Lw	-O-P-Q

Numbers in parentheses are mass numbers of most stable isotope of that element.

Many bonds that are covalent are inherently strong. For example, the energy of bond breaking, E_B (Kcal/mol), for common structures is as follows:

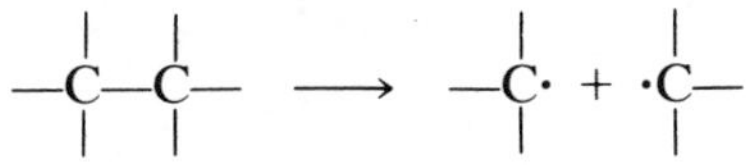

$$-\overset{|}{\underset{|}{C}}-\overset{|}{\underset{|}{C}}- \longrightarrow -\overset{|}{\underset{|}{C}}\cdot + \cdot\overset{|}{\underset{|}{C}}- \qquad E_B = 80\text{–}150 \text{ Kcal/mole}$$

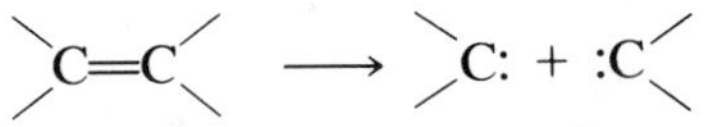

$$>C=C< \longrightarrow >C: + :C< \qquad \sim 150$$

$$-\overset{|}{\underset{|}{C}}-H \longrightarrow -\overset{|}{\underset{|}{C}}\cdot + \cdot H \qquad = 80\text{–}100$$

$$-\overset{|}{\underset{|}{C}}-O- \longrightarrow -\overset{|}{\underset{|}{C}}\cdot + \cdot O- \qquad \sim 80$$

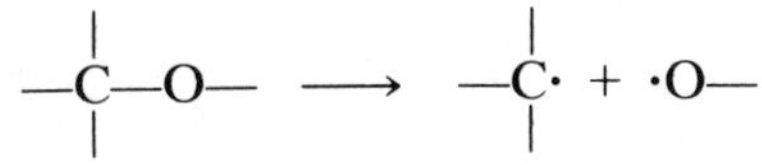

The important comparison is between the bond energy and the average thermal energy of RT or $\sim$ 0.6 Kcal/mol at room temperature. The ratio E_B/RT is about 200; hence, spontaneous bond breaking is infrequent ($\sim$ exp($-E/RT$) $\sim$ exp (-200) $\sim$ 10^{-60}).

Covalent *bond lengths* depend mainly on the two atoms involved and the number of bonds between them. The following are internuclear distances of some typical structures.

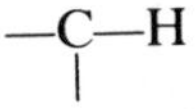

$-\underset{|}{C}-H$	1.08 ± 0.01A
$-\overset{|}{\underset{|}{C}}-\overset{|}{\underset{|}{C}}-$	1.53 ± 0.03
$>C=C<$	1.33 ± 0.02 (1.39 aromatic)
$-\overset{|}{\underset{|}{C}}-O-$	1.35–1.43
$>C=O$	1.20–1.25
$-\overset{|}{\underset{|}{C}}-N<$	1.47 (mostly)

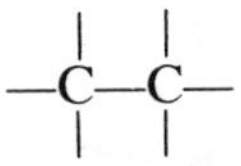

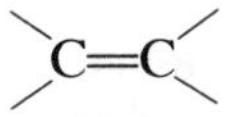

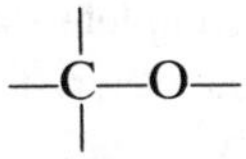

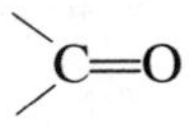

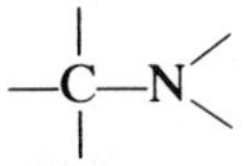

Covalent bond angles depend primarily on the central atom and its bonding. Some examples are as follows:

—C— (four single bonds)	tetrahedral	$\theta \sim 109°$
>C=	planar	$\theta \sim 120°$
>N—	pyramidal	$\theta \sim 107°$
—O\	bent	$\theta \sim 109°$

Rotational freedom around single bonds is illustrated in Figure 1-1. This figure provides a view along the bond.

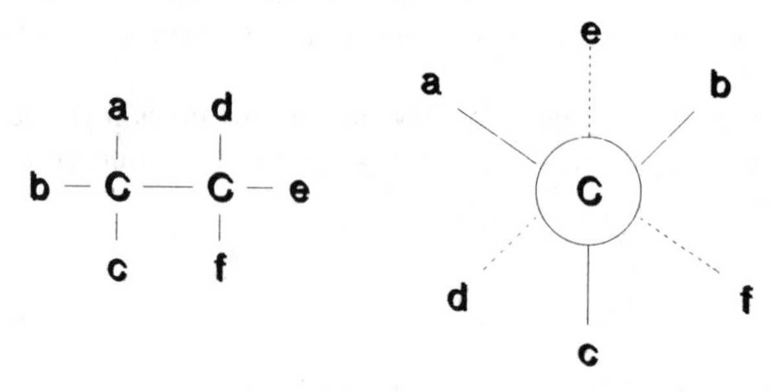

Figure 1-1 Illustrates the rotational freedom around single bonds

Atoms have positively charged centers (the atomic nuclei) surrounded by a negatively charged cloud (the electrons). The net charge is zero but covalent bonding can produce some charge displacement (complete transfer of electrons in the case of ionic bonding) from one atom toward the other. The bond is polar with one atom slightly positive and the other slightly negative (overall electrically neutral, however). The result is a permanent dipole. Some common examples are as follows:

—C—C— , —C—H	nonpolar
=C—C— , =C—H	weakly polar (=C negative)

$$-\overset{|}{\underset{|}{C}}-O- \quad , \quad -O-H$$

$$\rangle C{=}O \quad , \quad -\overset{|}{\underset{|}{C}}-Cl$$

polar
(O, N, Cl negative)

$$-\overset{|}{\underset{|}{C}}-N\langle \quad , \quad \rangle N-H$$

Because of bond polarities, organic molecules often have parts that are slightly charged electrically (either positively or negatively) although overall they are electrically neutral. These permanent electrical dipoles attract one another, so separate molecules tend to adhere. *Dipole-dipole* interactions occur for molecules that contain permanent dipoles.

Even molecules without permanent dipoles (nonpolar molecules) attract one another electrically through induced polarization. The electron clouds "bounce around" thermally, so instantaneously the molecules have positive and negative parts. These give rise to *induced dipole-induced dipole* interactions which are weaker than *dipole-dipole* interactions but also favor adherence (referred to as *dispersion forces*). There are also dipole-induced dipole attractions and a relatively strong interaction called hydrogen bonding. Intermolecular distance, melting temperatures, boiling temperatures, and heats of melting and evaporation depend on the strength of these nonbonded interactions and the molecular geometry (shape and "packability").

1.2 CONCEPTS IN MACROMOLECULAR STRUCTURE AND CHAIN CONFORMATION

In referring to *chain length*, it is important to note that synthetic polymers are polydisperse, that is, the polymerization products contain molecules with a range of sizes. Chemists often refer to chain-length distribution in describing polymeric structures. To the engineer and product development specialist, the more general term of *molecular weight distribution* has great significance. It is important to note that a distribution of sizes is a natural consequence of the random, statistical nature of all reactions at the molecular level. Hence, it is not a special situation for polymerization reactions. It is possible to predict this distribution in some cases and to control it to a certain degree.

The construction of long chains can place special requirements on the reaction. It must be essentially devoid of side reactions that interrupt chain growth. A reaction that yields 90 percent of the desired product, with only 10 percent side reactions, is often considered excellent in small-molecule chemistry, whereas in polymerization chemistry this might be a disaster. In fact one would whereas in polymerization chemistry this might be a disaster. In fact one would not even result in a respectable polymer (let's say more that 100 mers per molecule, on average) unless the linking yield is above 99 percent. This fact restricts our choices to a limited number of acceptable polymerization reactions.

If we examine structures from the viewpoint of large-scale architecture, then non-linear chains can be prepared, or branching reactions may be an inherent consequence of the reaction. Figure 1-2 illustrates the common forms of branching structures.

For copolymers, we may have a random distribution, block, or graft:

```
Random      AAABABBABBAABAB
Block       AAAAAAABBBBBBBB
Graft       AAAAAAAAAAAAAAAAAAAA
                B      B      B
                B      B      B
                B      B      B
                B      B      B
                B      B      B
```

Polymers may be further characterized in accordance with their macromolecular organization. In this regard there are crystalline and amorphous polymers. A crystalline polymer requires highly regular chains to achieve any crystallization at all. They melt over a range of temperatures up to some T_{max}. Crystalline regions are interspersed with disordered regions, and any chain has part of its length in both. This connected matrix of hard crystals, tied together by chains running through the softer disordered regions, confers a desirable mechanical toughness not possessed by small-molecule solids with the same chemical microstructure (a C_{20} wax is brittle, a $C_{1,000}$ polyethylene is tough).

Some polymers never crystallize, primarily because their chemical microstructure is too disorderly. Inherently amorphous polymers remain liquid-like in physical microstructure at all temperatures, but they always have a glass-transition temperature T_G. Amorphous polymers with a low T_G relative to ambient temperatures provide the basis of *synthetic rubbers*. Amorphous polymers with a high T_G (relative to ambient) find applications based on their glass-like behavior. No structural change occurs at T_G. The rate of local molecular rearrangement is the crucial feature for the glass transition.

When materials are in a crystalline state, individual chains acquire very regular conformations (planar zig-zag, helical, and so on) and pack together in very regular ways, thereby developing *long-range order* and a lower energy than in the liquid state.

The liquid state of a polymer refers to the amorphous regions in crystallized polymers, polymer melts, rubber, glasses, and polymer solutions. Most polymers of interest are flexible, which means they have more than one rotational state for the bonds along the chain backbone. These bonds can move freely among

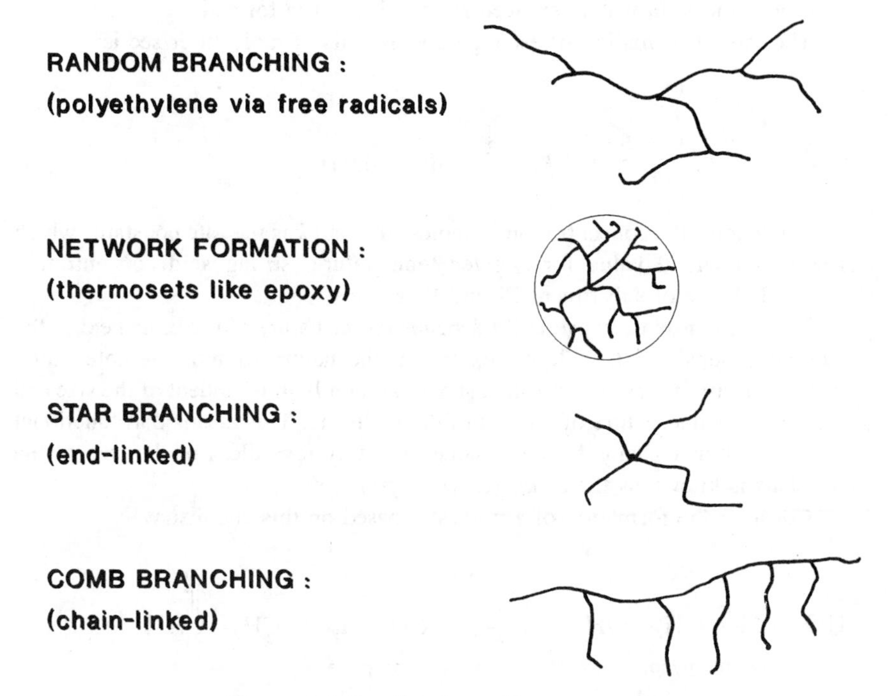

Figure 1-2 Possible branching structures

rotational states. Exceptions to this are glasses, where the chain conformations are essentially frozen. The chain units of different molecules pack in a liquid-like way, which means *short-range order* instead of long-range order. The individual chains have random conformations and are referred to as *random coils*.

In special cases of stiff chain polymers, the chains can develop long-range order along one direction but remain disordered along the other two directions. This is referred to as a *liquid crystalline state*.

1.3 POLYMERIZATION CHEMISTRY

The process of polymerization depends upon the reaction between functional groups. It is important to note that monomers must be capable of forming a least two links to other molecules. To examine how the reactivity of a functional group depends on the other substituents in the molecule and the overall molecular size, let us consider an ester-link formation:

$$R{-}\overset{\overset{\displaystyle O}{\|}}{C}{-}OH + R'OH \xrightarrow{\text{acid}} R{-}\overset{\overset{\displaystyle O}{\|}}{C}{-}OR' + H_2$$

The chemical equation describes the reaction between carboxylic acid functions and alcohol functions, forming ester linkages. The reaction is catalyzed by a strong acid with water removed as rapidly as it is formed.

The rate of formation of ester groups in units of moles/liter/sec is:

$$\frac{d[-\overset{\overset{\large O}{\|}}{C}-O-]}{dt} = k[-\overset{\overset{\large O}{\|}}{C}-OH][-OH]$$

where [ ] denotes the concentration in moles/liter and k is the rate constant, which has some numerical value for a given temperature, strong acid concentration [H+], and choice of substituent R- and R' groups.

The rate constant, k, primarily depends on the nature of the units next to the functional groups and to a lesser extent on the nature of more remote units. Hence, reactivity is locally controlled and is essentially independent of the size and detailed chemical structure of the rest of the molecule. This means that functional groups on polymer chains have the same reactivity regardless of the size of the chain. This is known as the *equal reactivity principle.*

Consider the formation of a polyester based on this chemistry

$$\underset{\text{ethylene glycol}}{HO-CH_2-CH_2-OH} + \underset{\text{adipic acid}}{HO-\overset{\overset{\large O}{\|}}{C}-CH_2CH_2CH_2CH_2-\overset{\overset{\large O}{\|}}{C}OH}$$

All molecules will have one of the following structures at any extent of reaction:

$$H-[O-(CH_2)_2-O-\overset{\overset{\large O}{\|}}{C}-(CH_2)_4-\overset{\overset{\large O}{\|}}{C}]_i-OH \qquad \text{alcohol, acid end}$$

$$H-[O-(CH_2)_2-O-\overset{\overset{\large O}{\|}}{C}-(CH_2)_4-\overset{\overset{\large O}{\|}}{C}]_i-O-(CH_2)_2-OH \qquad \text{alcohol, alcohol ends}$$

$$HO-\overset{\overset{\large O}{\|}}{C}-(CH_2)_4-\overset{\overset{\large O}{\|}}{C}[-O-(CH_2)_2-O-\overset{\overset{\large O}{\|}}{C}-(CH_2)_4-C]_i-OH \qquad \text{acid, acid ends}$$

where i is $0, 1, 2, 3, \ldots.$

Note that all unreacted acid or alcohol groups have the same local structure, regardless of i. Thus, the overall rate of reaction will slow down as the conversion proceeds:

($[-\overset{\overset{\displaystyle O}{\|}}{C}-OH]$ and $[-OH]$ decrease, so $d[-\overset{\overset{\displaystyle O}{\|}}{C}-O-]/dt$ gets smaller), but not because the reactivity decreases with increasing molecular size. It slows down similarly in nonpolymer ester formation (when monofunctional acids and alcohols are used). The same principle of equal reactivity applies in addition polymerization also. In the free-radical mechanism, the intrinsic reactivity of the radical species is independent of how long the growing chain has become.

The extent of any chemical reaction depends on two factors: the difference in energy between reactants and products (that is, the enthalpy change ΔH, or heat of reaction), and the difference in *entropy* (ΔS) between reactants and products. The thermodynamic quantity of great significance is ΔG, the change in *free energy* with reaction, which depends on ΔH, ΔS, and the absolute temperature T:

$$\Delta G = \Delta H - T\Delta S \tag{1-1}$$

The more negative Δ is, the more favorable the reaction is.

If the energy of the product molecules is lower than the energy of the reactant molecules, then ΔH is negative and heat is evolved as the reaction proceeds (the reaction is exothermic; the sum of bond energies is lower for the products than for the reactants). Hence, a negative Δ favors the reaction.

Entropy is very simply a measure of the *randomness* of a system and is related to the number of different ways the molecules in the system can be arranged. The Boltzmann equation relates Ω, the number of distinguishable configurations, to entropy S.

$$S = k \ 1n \ \Omega \tag{1-2}$$

This equation applies when all the configurations have the same energy. The Boltzmann constant k is equal to R/N_a, where R is the universal gas constant (8.33×10^7 ergs/°K/mole) and N_a is Avogadro's number (6.02×10^{23} moles/mole). The value of k in c.g.s. units is 1.38×10^{-16} ergs/°K/mole.

Entropy plays an important role. To illustrate, let's consider the following polymerization process:

$$nA \rightarrow A_n$$

or

$$nA + nB \rightarrow (AB)_n$$

where A and B are monomers, and A_n and $(AB)_n$ are polymer molecules with n repeating units. Schematically, this can be represented by Figure 1-3.

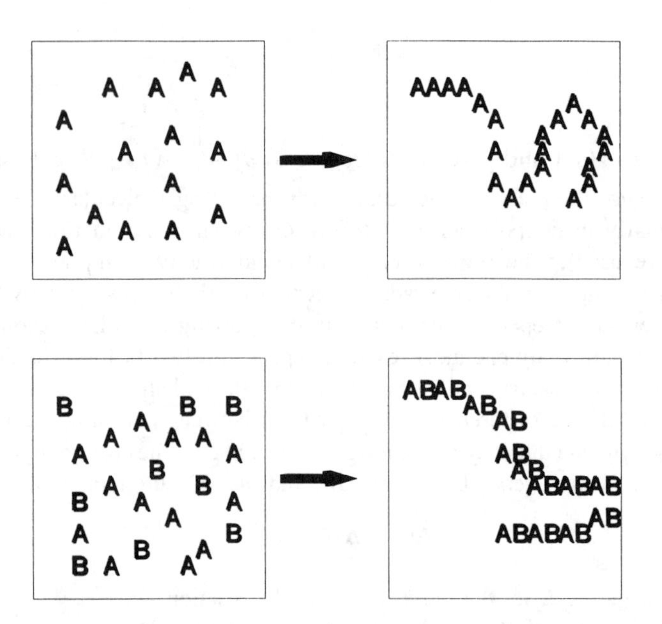

Figure 1-3 Pictoral view of example polymerization

During polymerization, entropy must decrease. Unconnected monomers have many more available arrangements than connected monomers, which must always remain together. Thus, ΔS is inherently negative for any monomer. Hence, $-T\Delta S$ is an inherently positive quantity for polymerization. Thus ΔS must be sufficiently negative (or exothermic) to result in a negative ΔG, in order for polymerization to occur.

Polymerization can occur only if ΔG is negative, so ΔH must be negative and large enough, *or* a by-product must be formed and removed to drive the reaction. This is known as LaChatlier's principle. There must also be a mechanism or chemical pathway that allows polymerization to occur at a reasonable rate. Just because a polymerization can occur does not mean that it will. Finally, since the unfavorable $T\Delta S$ part increases with temperature, every polymer has a *ceiling temperature* above which it cannot be formed. However, a polymer may be stable above its ceiling temperature if no mechanism for depolymerization is available.

Typically ΔH and ΔS are insensitive to temperature. Also, ΔS per mole of monomer polymerized is remarkably constant for many monomers and is around -26 cal/°K/mole. Thus, to polymerize (ΔG less than zero) at 300°K, the heat of polymerization must be at least (300)(-26) $\sim$ -7.8 kcal per mole of monomer.

The basic requirements to achieve long chains are:

- difunctional monomers
- reaction proceeding nearly to completion

- absence of side reactions that produce unreactive chain ends or rings
- very accurate control of reactant stoichiometry

Some comments on ring formation are warranted at this point. Since five-member and six-member rings close easily, monomers must be selected to avoid this possibility. Smaller rings are highly strained, as are seven- to fourteen-member rings. Larger rings are "favorable" but not very probable. Nylon 2,2 would have ring-forming problems, for example, because it can form a six-member ring:

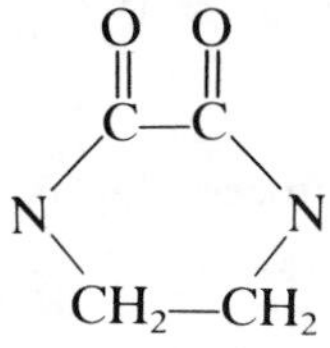

```
      O   O
      ||  ||
      C — C
     /     \
    N       N
     \     /
    CH2—CH2
```

Note that the smallest ring for Nylon 6,6 has 14 members. In ordinary undiluted condensations, linear polymers are obtained if the smallest ring is seven or eight. Thus, Nylon 6 is obtained from ring opening of the seven-member caprolactam monomer.

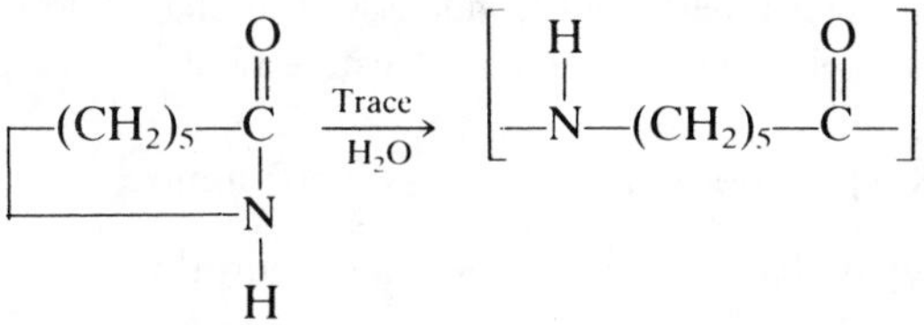

```
              O                      H              O
              ||       Trace         |              ||
  ┌—(CH2)5—C      ————————→    [ —N—(CH2)5—C— ]
  |           |        H2O
  └———————N
              |
              H
```

Nylon 4 or 5 would be hard to achieve, however.

At high temperatures, unreacted ends can exchange with already formed linkages, but that does not affect the degree of polymerization.

Now it is important to note that there are several types of polycondensations. The first of these worth noting is high-temperature bulk. In this case, the inherent ΔG is small (reversible) but driven to completion by the removal of a volatile product. For example:

$$\text{acid} + \text{alcohol} \longrightarrow H_2$$

$$\text{acid} + \text{amine} \longrightarrow \text{polyamide} + H_2O$$

Also ester interchange:

$$CH_3O-\overset{O}{\overset{\|}{C}}-C_6H_4-\overset{O}{\overset{\|}{C}}-OCH_3 + HO-CH_2CH_2OH \longrightarrow$$

dimethyl terephthalate ethylene glycol

$$CH_3OH\uparrow + \left[-\overset{O}{\overset{\|}{C}}-C_6H_4-\overset{O}{\overset{\|}{C}}-O-(CH_2)_2-O-\right]$$

methanol

As an example, the commercial process for polyethylene terephthalate involves the following two considerations:

- Methanol is more volatile than water so the reaction proceeds at lower temperatures.
- The ester monomer is more soluble in glycol than the acid monomer, so contact is better.

The second important type of polycondensation is that of low-temperature solution. Here, the inherent ΔG is large and negative (that is, irreversible):

acid chloride + amine $\longrightarrow$ polyamide + HCl↑

isocyanate + alcohol $\longrightarrow$ polyurethane

phenol + phosgene $\longrightarrow$ polycarbonate + HCl↑

chlorosilane + water $\longrightarrow$ silicone + HCl↑

$$Cl-\overset{O}{\overset{\|}{C}}-(CH_2)_8-\overset{O}{\overset{\|}{C}}-Cl + H_2N-(CH_2)_6-NH_2 \longrightarrow HCl\uparrow + \text{Nylon 6,10}$$

$$O{=}C{=}N-(CH_2)_6-N{=}C{=}O + HO-(CH_2)_2-OH \longrightarrow$$

$$\left[-O-\overset{O}{\overset{\|}{C}}-\overset{H}{\overset{|}{N}}-(CH_2)_6-\overset{H}{\overset{|}{N}}-\overset{O}{\overset{\|}{C}}-(CH_2)_2-\right]$$

$$HO-C_6H_4-\underset{CH_3}{\overset{CH_3}{\underset{|}{\overset{|}{C}}}}-C_6H_4-OH + Cl-\overset{O}{\overset{\|}{C}}-Cl \longrightarrow HCl\uparrow +$$

$$\left[O-\overset{O}{\overset{\|}{C}}-O-C_6H_4-\underset{CH_3}{\overset{CH_3}{\underset{|}{\overset{|}{C}}}}-C_6H_4\right]$$

$$Cl-\underset{CH_3}{\overset{CH_3}{\underset{|}{\overset{|}{Si}}}}-Cl + H_2O \longrightarrow \left[O-\underset{CH_3}{\overset{CH_3}{\underset{|}{\overset{|}{Si}}}}\right] + HCl\uparrow$$

Other examples are epoxys, phenol-formaldehyde and urea-formaldehyde resins.

The characteristics of this type of polycondensation are:

- rapid reaction but good heat transfer
- low-temperature advantage
- dilution promotes ring formation
- solvent must be removed afterward

Linear polycondensations result in a distribution of chain lengths. Some molecules will have many repeat units, while others will have less. In practice, *the molecular weight distribution* is characterized by average molecular weights that can be determined experimentally. Defining P_r as the fraction of polymer molecules in the system with r mers, probability theory can be used to define important modes of the molecular weight distribution. For a linear polycondensation of *A-A* monomers with *B-B* monomers, forming *AB* linkages, assume there are equal moles of *A* and *B* functions in the system which react to an extent p, where p is the fraction of *A* (or *B*) functions reacted. Assuming all molecules have unreacted functions on their ends, we may apply the equal reactivity principle. P_r is defined as the probability that if a molecule is selected at random, it will contain exactly r mers. Each molecule has two unreacted functions, so if an unreacted function is selected at random, we will have picked a polymer molecule without prejudice. Therefore, the question is: How big is the molecule? We shall assume the following example:

$$A - AB - BA - AB \cdot BA - AB\,BA - AB \cdot B$$

For the simple case of A - A or B - B,

$$P_1 = 1 - p$$

The only way that a chain with one mer is obtainable is if the attached functional group is unreacted. The probability of that is 1-p, which is the fraction of unreacted functions in the system. Likewise,

$$P_2 = p(1 - p) \qquad A - AB - B \quad \text{or} \quad B - BA - A$$

$$P_3 = p^2(1 - p) \qquad A - AB - BA - A \quad \text{or} \quad B - BA - AB - B$$

or, in general,

$$P_r = p^{r-1}(1 - p)$$

Note that $P_1 + P_2 + P_3 + \sum_{r=1}^{\infty} P_r = 1$. In other words, the sum of all probabilities (sum of all fractions of molecular size) must be unity. Hence,

$$\sum_{r=1}^{\infty} P_r = (1 - p)(1 + p + p^2 + \ldots)$$

$$= \frac{1 - p}{1 - p} = 1 \qquad \text{(ck.)}$$

As a check:

$$\frac{(1)P_1 + (2)P_2 + 3(P_3) \ldots}{P_1 + P_2 + P_3 \ldots} = \frac{\text{moles of mers}}{\text{moles of molecules}}$$

$$\overline{DP} = \frac{\sum rP_r}{\sum P_r} = \frac{1 - p}{(1 - p)^2} = \frac{1}{1 - p}$$

Where $\overline{DP}$ = average degree of polymerization.

Figure 1-4 shows the probability distribution function.

Note that $r = 1$ is the most frequent molecular size even when the degree of polymerization is large. When the degree of polymerization (DP) is large, the distribution can be stated as:

$$P_r = \exp[-r/a] = /a$$

where $a = DP$. This expression is referred to as the *exponential* distribution or *most-probable* distribution, which is quite important to polymerization.

Lets now consider P_r, and the average molecular weights under more general cases:

$$\overline{DP} \longrightarrow \overline{DP}_n = \frac{\sum rP_r}{\sum P_r} \qquad \text{(number-average } DP\text{)}$$

$$\overline{DP}_w = \frac{\sum r^2 P_r}{\sum rP_r} \qquad \text{(weight-average } DP\text{)}$$

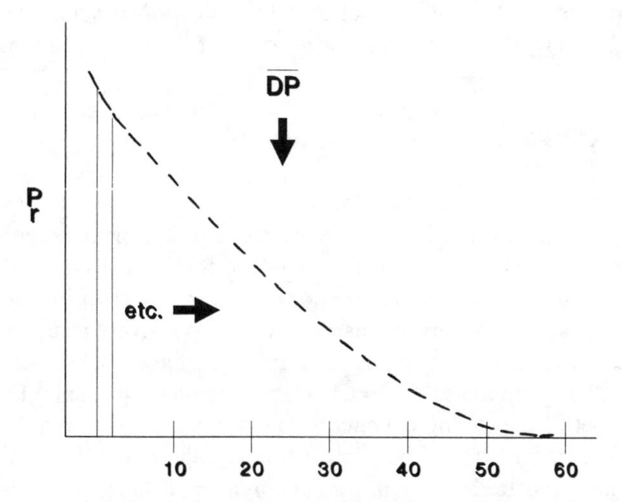

Figure 1-4 Probability distribution function

Defining m_o as the mer molecular weight, then

$$M_n = m_o DP_n \qquad \text{(number-average molecular weight)}$$

$$M_w = m_o DP_w \qquad \text{(weight-average molecular weight)}$$

P_r *P(M)* are number distributions (fractional number); W_r, *W(M)* are weight distributions (fractional weight).

$$W_r = \frac{rP_r}{\sum rP_r}$$

from whence

$$W(M)dM = \frac{MP(M)dM}{\int_0^\infty MP(M)dM} \qquad \text{(continuous form)}$$

W(M)dM equals the fractional weight of the polymer contributed by molecules with molecular weights between *M* and $M + dM$.

Some methods provide values of M_n and others provide M_w. M_w/M_n is defined as the most probable distribution.

$$\frac{\overline{M_w}}{\overline{M_n}} = m_0 \frac{(\sum r^2 P_r(\sum P_r)}{(\sum r P_r)^2} = 2 \qquad \text{for } \overline{DP_n} \text{ large}$$

Mw/Mn = 1 for monodisperse polymers and is much greater than 2 for many commercial polymers. This ratio is referred to as the *polydispersity* or *dispersity ratio*. M_w/M_n is essentially a convenient measure of the distribution breadth of a polymer.

1.4 CHAIN-GROWTH POLYMERIZATION

We now direct attention to the subject of chain growth polymerization; also known as *addition polymerization*. Chains grow by the addition of monomers to active species that are created in the system. There are three main types: *free-radical* polymerization, *cationic* polymerization, or *anionic* polymerization, according to the nature of the active species (uncharged radical, positively charged cation, or negatively charged anion). The monomers involved have double bonds, mostly vinyl $H_2C{=}CHR$, vinylene $H_2C{=}CHR'$, or conjugated diene $H_2C{=}CHR{-}HC{=}CH_2$. A set of competing chemical reactions is involved, and their relative rates control the overall rate of polymerization and the molecular weights of the polymer that is synthesized. For commercial polymers, the most important of these reactions are as follows:

1. Initiation—creation of active species
2. Propagation—incorporation of monomer into the growing chain
3. Termination—destruction of active species
4. Transfer—transfer of an active species from one molecule to another

During initiation, the reaction forms pairs of free radicals.

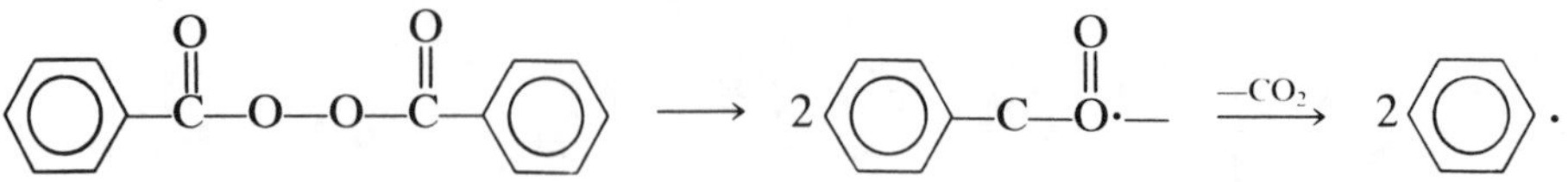

$$C_6H_5{-}\overset{O}{\overset{\|}{C}}{-}O{-}O{-}\overset{O}{\overset{\|}{C}}{-}C_6H_5 \longrightarrow 2\,C_6H_5{-}\overset{O}{\overset{\|}{C}}{-}O\cdot \xrightarrow{-CO_2} 2\,C_6H_5\cdot$$

benzoyl peroxide (BPO)

$$CH_3{-}\underset{}{\overset{C\equiv N}{\overset{|}{C}}}{-}N{=}N{-}\overset{C\equiv N}{\overset{|}{C}}{-}CH_3 \longrightarrow \xrightarrow{-N_2} 2\,CH_3{-}\underset{CH_3}{\underset{|}{\overset{C\equiv N}{\overset{|}{C}}}}\cdot$$

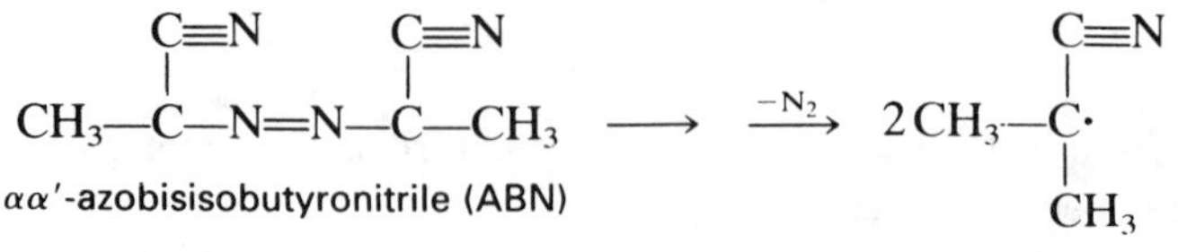

$\alpha\alpha'$-azobisisobutyronitrile (ABN)

Both BPO and ABN are stable at room temperature, but both decompose slowly at elevated temperatures and are good initiators at 60-90°C. Other peroxides are useful at other temperatures. *Small amounts* are dissolved in the monomer or monomer solution; thermal decomposition provides a steady supply of free radicals to the system.(ABN is also a photoinitiator, providing radicals by exposure to UV light at room temperature and below). Any source of free radicals works, including H_2O_2 + Fe^{+2} in aqueous monomer solution.

During *propagation*, free radical adds to the monomer double bonds:

$$R\cdot + (H)(H)C{=}C(H)(R') \longrightarrow R{-}CH_2{-}\dot{C}(H)(R')$$

$$R{-}CH_2{-}\dot{C}(H)(R') + CH_2{=}C(H)(R') \longrightarrow RCH_2CHR{-}CH_2{-}\dot{C}(H)(R)$$

During *termination*, radicals react to form stable molecules

$$R{-}CH_2{-}\dot{C}(H)(R') + R{-}CH_2{-}\dot{C}(H)(R') \longrightarrow R{-}CH_2{-}CH(R'){-}CH(R'){-}CH_2{-}R \quad \text{(called } \textit{coupling}\text{)}$$

or

$$\longrightarrow R{-}CH{=}C(H)(R') + H{-}CH(R'){-}C(H)(H) \quad \text{(called } \textit{disproportionation}\text{)}$$

During *transfer*,

$$R{-}CH_2{-}\dot{C}(H)(R') + (H)(H)C{=}C(H)(R')$$

$$\longrightarrow R{-}CH_2{-}CH(R'){-}H + (H)(H)C{=}\dot{C}(R')$$

The transfer is to the *monomer*, not necessarily H, but it could be an atom or group in R'.

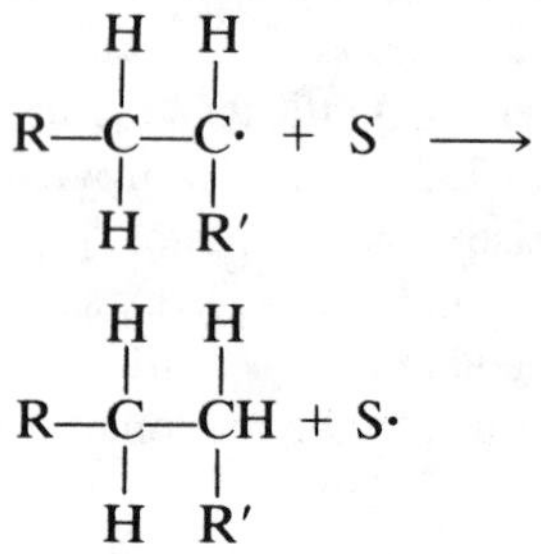

```
     H  H
     |  |
  R—C—C· + S  ⟶
     |  |
     H  R'

     H  H
     |  |
  R—C—CH + S·
     |  |
     H  R'
```

This shows transfer to a solvent, again, not necessarily H.

Growth-interruption reactions (termination and transfer) compete with the propagation reaction at each step in the growth of chains in free-radical polymerization. Thus, interruption could occur after one monomer addition, or two, or three, and so on, and the probability of interruption is the same at each step. The probability of interruption after four steps (to form an "inactive" chain) is governed by the same law as condensation polymerization, although the mechanism is quite different:

$$Pr = \exp(-r/a)/a \qquad (a \equiv \text{large})$$

where $a = DP_n$, as before. If no coupling occurs (termination by disproportionation, solvent transfer, and monomer transfer, or some combination), the most probable distribution for the chains being formed at each moment is obtained. The basis for this concept is steady state. Fully grown polymer molecules are synthesized practically from the beginning of the polymerization process. The value of a is DP_n for the polymer being formed at the particular instant of time being considered, so the preceding expression describes the *instantaneous* number distribution.

The weight distribution, $rP_r/\sum rP_r$, is

$$Wr = r\exp(-r/a)/a^2$$

or

$$W(M)dM = M\exp(-M/\bar{M}_n^2)dM$$

where *W(M)* is the weight distribution of molecular weights ($\overline{M_n} = m_o\overline{DP_n}$). Thus, as previously

$$\frac{\bar{M}_w}{\bar{M}_n} = \frac{\overline{DP}_w}{\overline{DP}_n} = 2$$

If termination is by coupling only, with no transfer, the distribution is narrower:

$$P_r = r \exp(-r/a)/a^2$$

where now $DP_n = 2a$, and

$$\frac{\bar{M}_w}{\bar{M}_n} = \frac{\overline{DP}_w}{\overline{DP}_n} = \frac{3}{2}$$

As conversion proceeds, the distribution of chains being formed at each moment may change as the conditions change:

$$\frac{1}{\overline{DP}_n} = \frac{K[I]^{1/2}}{[M]} - + C_M + \frac{C^S[S]}{[M]}$$

Thus, the polymer at the end is the accumulation of chains formed under different conditions, and the accumulated distribution broadens with conversion. Values of $\overline{M_w}/\overline{M_n}$ are very commonly in the range of 2 to 3 unless polymer transfer occurs. Polymer transfer gives long chain branching and can produce very broad distributions ($\overline{M_w}/\overline{M_n}$ up to 20 or even more). Free-radical polymerization of ethylene (giving LDPE, or low-density polyethylene) is the classical example, but it can happen in other vinyl polymers also (polyvinyl acetate).

Other important features to bear in mind about high conversions are:

- Viscosity increases, heat removal becomes more difficult, so the reaction may "run away."
- Viscosity slows down termination reactions, radical concentration builds up, so the reaction can run away again.
- The polymer may be insoluble in its own monomer (vinyl chloride, acrylonitrile), which also can suppress termination, so again the reaction can run away.

With cationic polymerization, initiation is achieved through the use of strong acids such as H_2SO_4, $AlCl_3$, BF_3, usually with coinitiators such as H_2O, ROH, RCl, and ethers. The propagating species is a carbenium ion - $C^{\oplus}$with the counterion (negatively charged group) in close proximity. The rate of propagation strongly depends on the counterion and the choice of solvent. Cationic polymerization is highly susceptible to transfer reactions, usually run at low temperature to minimize these and other side reactions. Furthermore, it is easily poisoned by polar impurities and must be run under scrupulously clean conditions.

The kinetics of cationic polymerization are complex and uncertain compared with free radical polymerizations. They may have rates of conversion that decrease with increasing temperature. It is a type of polymerization that works with only a few monomers; for example, vinyl ethers, styrene, isobutylene, and aldehydes.

1.5 SOLUTION PROPERTIES

In this section we will discuss general properties and characteristics of polymer solutions. To begin, crystalline polymers are usually only soluble near and above their melting temperatures. For examples, polyethylene (T_m = 130°C) is not soluble below about 80°C, but it is soluble in many solvents above that range. Nylon (T_m = 230°C) is soluble at room temperature, but only in strongly hydrogen-bonding solvents.

In contrast, amorphous polymers are soluble in many solvents. It's important to note, however, that molecular weight has very little effect on solubility, although the rate of dissolution can be very slow for high molecular weight polymers. The terms *good solvents*, *poor solvents*, *nonsolvents*, and *theta solvents* are applied depending on the molecular interactions of polymer units and solvent molecules.

Polymer networks can imbibe (that is, absorb) solvents and swell, sometimes to many times their original volume, but they do not dissolve. The mutual solubility of different polymer species is very limited, except in special cases.

Thermodynamics essentially govern solubility; most notably a negative ΔG favors miscibility.

Solutions are more random than pure components, so ΔS is inherently positive, and thus the -TΔS term always favors the solution state.

The magnitude and sign of ΔH is governed by the *change* in intermolecular energy when a solution is formed. For *gases* the molecules are far apart, so intermolecular energies are small, and the change with mixing is very small. Entropy always wins $\Delta G < 0$, and hence all gases are miscible. For *liquids* the molecules are close, so intermolecular energies are significant. However, the *change* with mixing may be large or small depending on the like-dissolves-like principle. Either energy or entropy might win out. Many liquid pairs are miscible; others are not, and temperature is important. For *solids* the molecules are closely packed, so intermolecular energies are large. Also, geometry and packing arrangement are important. Miscible solids (solid solutions) are rare in nature. For a *solid plus a liquid*, there is always a solubility limit.

An important question to ask is why polymers differ in solubility behavior. To understand the answer to this it is important to recognize that the entropy of mixing per polymer molecule is about the same as that for small molecules. The energy change per polymer molecule, however, increases in direct proportion to chain length (number of mers, *DP*). Even a small energy penalty *per mer* becomes a large ΔH *per polymer molecule* if *DP* is large. For long chains, ΔH can overwhelm the always favorable ΔS even in closely matched "like-dissolves-like" components, such as polyethylene and polypropylene. This leads us to the Flory-Huggins model which is based on regular solution concepts.

$$\Delta S = k \ln \frac{\Omega(\text{soln})}{\Omega(\text{pure})}$$

where Ω is the number of distinguishable configurations.

To appreciate what happens during mixing, examine Figure 1-5. Three cases are illustrated: (A) α solution made by the mixing of small molecules, (B) a solution made by mixing small molecules and chain-like molecules, and (C) a mixture produced from only chain-like molecules.

Each molecule of the solvent, or mer of a polymer chain, occupies one site of volume v on a lattice. If the total volume of two pure components is V_1 and V_2, the number of sites in each is $V_1/^v = N_1$ and $V_2/^v = N_2$, respectively. If the components are mixed to form a solution, the solution will have $N_1 + N_2$ sites, each of volume v (assume no volume change on mixing), and the solution volume will be $V_1 + V_2$.

If we consider both components as small molecules, then the number of distinguishable configurations Ω is:

$$\Omega_1 = \Omega_2 = 1$$

$$\Omega_{1+2} = \frac{(N_1 + N_2)!}{N_1!\, N_2!}$$

and using $\Delta S = k[1n\Omega_{12} - 1n\Omega_2]$ and Stirling's approximation: $1nN! = N1nN - N$, we obtain:

$$\Delta S = k\left[N_1 \ln \frac{N_1 + N_2}{N_1} + N_2 \ln \frac{N_1 + N_2}{N_2}\right]$$

or

$$\Delta S = -\frac{(V_1 + V_2)}{V} R[\phi_1 \ln \phi_1 + \phi_2 \ln \phi_2]$$

where V is the volume per *mole* of sites (molar volume of components) and $R = k\,N_A$ is the universal gas constant (N_A is Avogadro's number). ($R/V = k/v$.) The volume fractions of the components are ϕ_1 and ϕ_2, respectively.

In both pure state and solution, each molecule at any moment interacts directly only with 6-12 nearest neighbors (call this lattice coordination number z). For each pure component, all nearest neighbors are like molecules, and the intermolecular energy is either $\epsilon_{11}z/2$ or $\epsilon_{22}z/2$ per lattice site. For the solution state, some of the 1-1 and 2-2 contacts are replaced by 1-2 contacts:

$$(1\text{-}1) + (2\text{-}2) \rightarrow 2(1\text{-}2)$$

(A) MIXTURES OF SMALL MOLECULES :

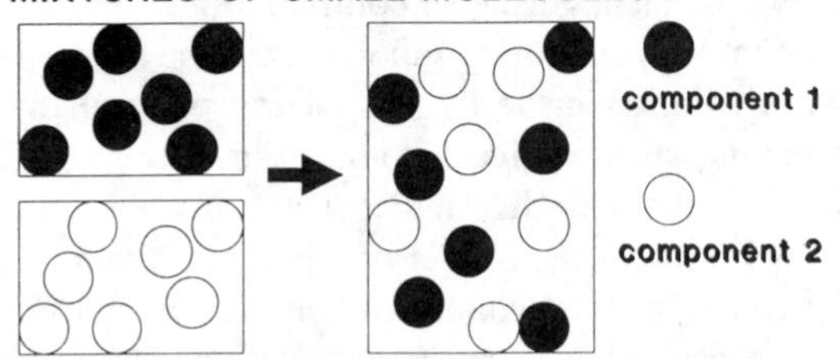

(B) MIXTURES OF SMALL AND CHAIN MOLECULES :

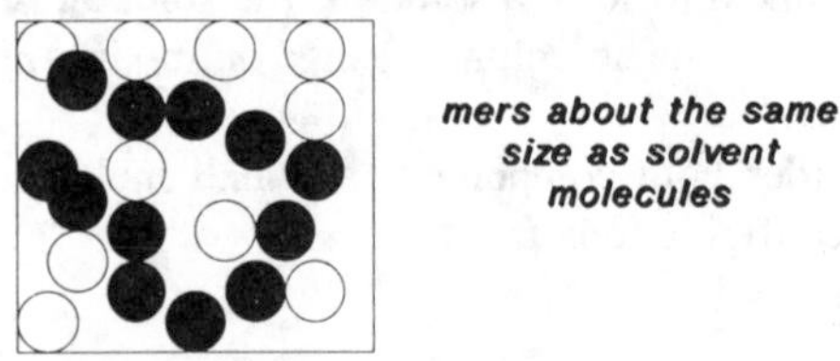

(C) MIXTURES OF CHAIN MOLECULES :

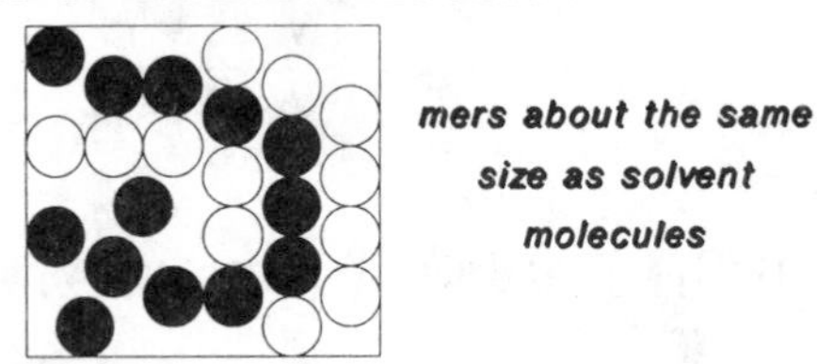

Figure 1-5 Illustration of mixture on a macroscopic scale

The energy change per dissimilar contact is

$$\Delta\varepsilon = \varepsilon_{12} - (\varepsilon_{11} + \varepsilon_{22})/2$$

and the number of such (nearest neighbor) dissimilar contacts is

$$zN_1\left[\frac{N_2}{N_1 + N_2}\right] \qquad \text{(random mixture)}$$

Thus, the enthalpy (energy) change when V_1 and V_2 volumes of pure components are mixed randomly is

$$\Delta H = (n_1 + N_2) z \Delta\varepsilon \frac{N_1 N_2}{(N_1 + N_2)^2}$$

or

$$\Delta H = \frac{V_1 + V_2}{V} RT\chi\phi_1\phi_2$$

where $\chi \equiv z\Delta\varepsilon / kT$ is called the interaction parameter, and ϕ_1 and ϕ_2 are the volume fractions, as before. Combining with the expression for ΔS:

$$\Delta G = \frac{V_1 + V_2}{\underline{V}} RT[\phi_1 \ln \phi_1 + \phi_2 \ln \phi_2 + \chi\phi_1\phi_2]$$

This is the expression for free-energy change when both components are small molecules (each occupies one lattice site). The same calculations can be carried through for polymers (each molecule occupies r lattice sites, where r is the number of mers per chain):

$$\Delta G = \frac{V_1 + V_2}{V} RT\left[\phi_1 \ln \phi_1 + \frac{\phi_2}{r_2} \ln \phi_2 + \chi\phi_1\phi_2\right]$$

$$\Delta G = \frac{V_1 + V_2}{V} RT\left[\frac{\phi_1 \ln \phi_1}{r_1} + \frac{\phi_2}{r_2} + \chi\phi_1\phi_2\right]$$

Note that the first two terms in brackets are inherently negative (coming from ΔS) and favor solution. The third term is almost always positive, meaning there is an energy penalty in forming the solution:

$$\chi = \frac{V}{RT}(\delta_2 - \delta_1)^2$$

where δ_1 and δ_2 are the *cohesive energy density* values for the pure components.

$$\delta = \left(\frac{E}{V}\right)^{1/2}$$

where E is the intermolecular energy per mole holding the molecules to one another, and V is the molar volume. Empirical values of δ are tabulated for many polymers and solvents. They represent the idea of like-dissolves-like in quantitative form. Specific interactions between components (acid-base, hydrogen bonding, and so on) can, however, give negative x and thereby, like entropy, also favor mixing.

Note that molecular size (large r) reduces the entropy contribution relative to the energy term, and especially so for mixtures of two polymers: x is very small for polyethylene and polypropylene units, but they still have positive ΔG if their chain lengths are long enough.

Polymer-solvent systems can be described in terms of phase diagrams (refer to Figure 1-6). In this diagram, T_c is the critical temperature. The skewness of the diagram increases with molecule size r. T_c increases with r to a maximum value ($r \rightarrow \infty$) called θ, known as the theta temperature, or the Flory temperature. The θ-value is obtained for any particular polymer-solvent system by constructing a plot like Figure 1-6B.

If $T >> \theta$, we have a *good solvent.*
If $T > \theta$, we have a *poor solvent.*
If $T = \theta$, we have a *theta solvent* (more accurately, a theta condition).
If $T << \theta$, we have a *non-solvent.*

The smaller the difference in cohesive energy density, δ_2-δ_1, the "better" the solvent.

Polymer-solvent thermodynamics are important for several reasons. First, the size of polymer molecules in a dilute solution (the volume each chain pervaded) depends on solvent "goodness." Applications that use polymers in solution, for example, lubricants, various oil field chemicals, drag reducers, protective coatings, and so on, depend on *chain dimensions* and thermodynamic interactions. A second consideration is that many polymerizations are performed in solution (for example, ethylene-propylene copolymer, polyisobutylene). Phase separation can occur at some stage in the conversion, which can affect the process. Also the finishing operation (solvent removal) depends on solvent-polymer interactions. Finally, the macromolecular structure determination (molecular weight, distribution, branching) is done in a dilute solution by a variety of techniques, and interpretation depends on the thermodynamic properties.

Of course, polymer-polymer thermodynamics influence the entire field of polymer blends and provide the basis for molecular design of block and graft copolymers as blend compatibilizers.

1.6 CHAIN CONFORMATION

Many polymers can be described as being "flexible," which means that rotation about the backbone bonds, if they are single bonds, is possible. Bond lengths and bond angles are fixed, but different rotational states have only small energy differences ($\Delta\epsilon \sim kT$). Even the intramolecular barriers to rotation are small enough that the bonds can "flip" between rotational states very rapidly in solution and melts. For polymers below the glass-transition temperature this flipping is

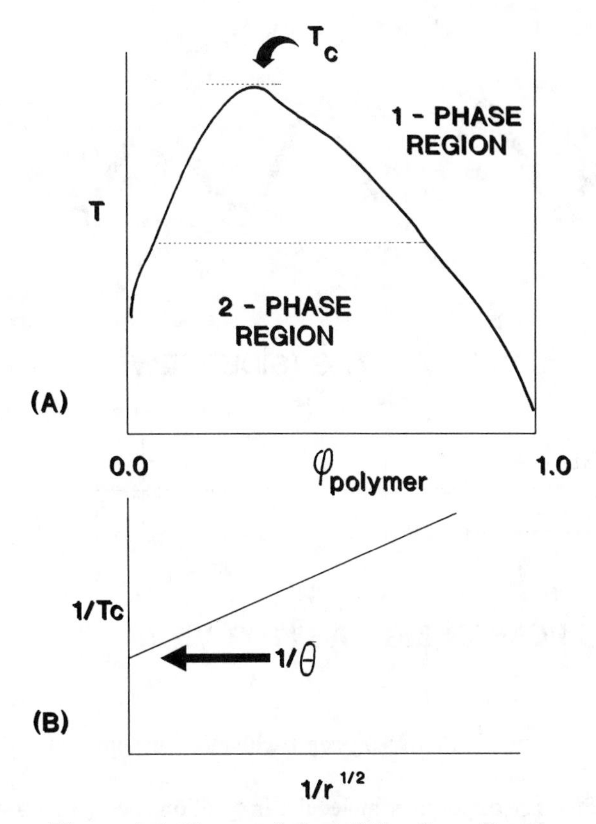

Figure 1-6 Characteristics of the phase diagram

negligible, however, and in any crystalline region the chains are fixed in very regular conformations by intermolecular forces. Figure 1-7 illustrates side and top views of the planar zig-zag conformation.

The planar zig-zag *backbone conformation* is a way of visualizing the *steric configuration* of a chain, which is fixed, but it is only one of many possible backbone conformations available.

There is an almost infinite number of conformations. If, for example, there are three rotational states for each bond (the three rotational energy minima), then

$$\text{number of conformations} = 3^{n-2}$$

so for a modestly long chain, $n = 1{,}000$ there are $3^{998} \sim 10^{475}$ different conformations, of which the planar zig-zag is only *one*. Even for a mole of such chains (6×10^{23}), the chances that any two have the same conformation is negligible. This just helps to illustrate the complexity of polymer structures.

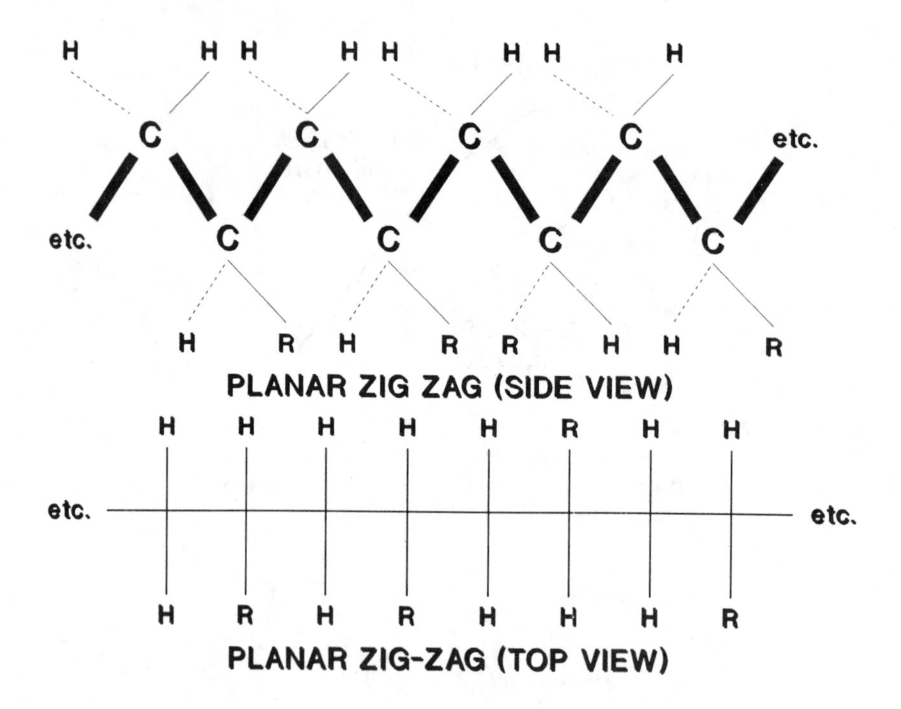

Figure 1-7 Planar zig-zag backbone configuration

The average *size* of polymer molecules in solution is very important, and one measure of size is the average end-to-end distance. A simplified model of size can be devised from random-walk theory. In the *random-walk* model, the chain backbone is a connected sequence of n bonds, each of length ℓ (ℓ = 1.53A for a C-C single bond). Each bond can choose any direction in space, regardless of the directions of the other bonds, and

ℓ = bond length (step size)
n = number of bonds (number of steps)
$n1$ = total chain length (length of the walk)

Label the steps from one end to the other $(1,2,3,\ldots,i,\ldots,n)$, and let a_i be the bond vector. Then

$$\vec{r} = \sum_{i=1}^{n} \vec{a}_i$$

where r is the vector from one end of the chain to the other (the random walk). One way to calculate the *average length of* r for a large number of such chains is to take the dot product of r with itself for each chain (this gives the square of the length of r),

$$r^2 = \vec{r} \cdot \vec{r} = \sum_{i=1}^{n} \sum_{j=1}^{n} \vec{a}_i \cdot \vec{a}_j$$

and then take the average of r^2 over all the chains

$$\langle r^2 \rangle = \sum_{i=1}^{n} \sum_{j=1}^{n} \langle \vec{a}_i \cdot \vec{a}_j \rangle$$

Hence, we need to evaluate $\vec{a}_i \cdot \vec{aj}>$ and then take the sum. But we know something about these terms.

If $i = j$ $\qquad \vec{a}_i \cdot \vec{a}_j = l^2$, so $\langle \vec{a}_1 \cdot \vec{a}_j \rangle = l^2$

If $i \neq j$ $\qquad \langle \vec{a}_i \cdot \vec{a}_j \rangle = 0$ (random directions)

The sum contains n^2 terms, but n of them have $i = j$ and the n^2 - n others have $i = j$. Thus,

$$\langle r^2 \rangle = nl^2 \qquad (\langle r^2 \rangle^{1/2} \propto M^{1/2})$$

Other properties that measure the mean chain dimensions can also be calculated. A particularly valuable one (measurable by a light scattering is the radius of gyration R_G; refer to Figure 1-8).

R^2G = mean square distance of chain units from the center of gravity of the chain.

Two major factors determine the dimensions of polymer chains. The first has to do with *local restrictions on chain conformation*. These depend directly on the chemical microstructure of the polymer and always act to expand the dimensions. Bond angles are fixed, so the chain can never be modeled as a completely unrestricted random walk. Also, rotational states are subject to steric restrictions which will depend on side group sizes and shapes, and so on.

The second factor concerns *long-range restrictions on chain conformation.* These depend on the thermodynamic interactions between the solvent molecules and polymer segments. A polymer chain differs from a random walk because it cannot intersect and cross over itself. The segments occupy some volume and no two segments can occupy the same space simultaneously. This is known as the

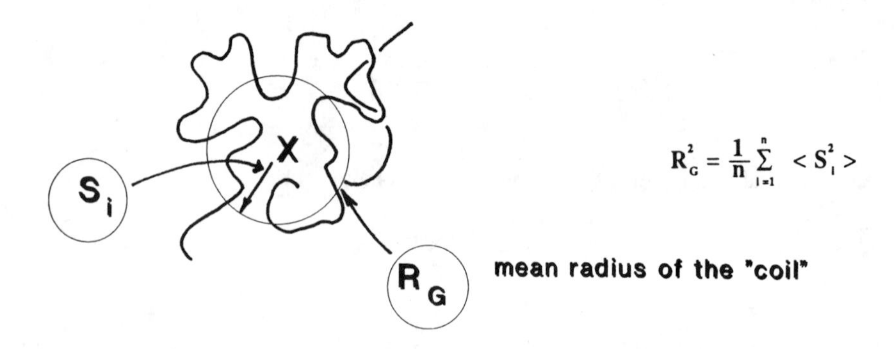

FOR UNRESTRICTED RANDOM WALKS

$$R_G^2 = \frac{nl^2}{6} = \frac{\langle r^2 \rangle}{6}$$

Figure 1-8 Radius of gyration of random walk model

excluded-volume effect, which is solvent dependent. Its effect is large in good solvents, weaker in poor solvents, and is exactly canceled in a theta solvent.

The behavior of the radius of gyration over the full range of molecular weights looks like Figure 1-9.

Poor solvents show intermediate behavior, and the polymer precipitates if the temperature gets much below . Also, in good solvents ($T >> \theta$), R_G is quite insensitive to temperature. The excluded-volume phenomenon is difficult theoretically. The best theoretical value for the exponent ($R_G \infty M^b$) is $b = 0.588$. Flory's early theory gave $b = 3/5 = 0.600$.

In general, R_G is smaller for branched chains than linear chains of the same total molecular weight.

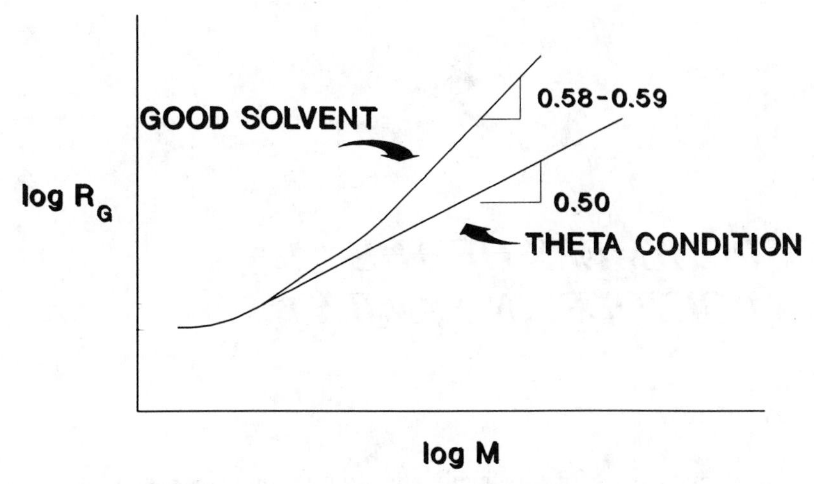

Figure 1-9 Radius of gyration as a function of molecular weight

2 *POLYMERS OF THE POLYOLEFIN FAMILY*

2.1 OVERVIEW

The principle monomers for this family are ethylene, propylene, and butadiene. The four major polyolefins are polyethylene, polypropylene, polyvinyl chloride (PVC), and polystyrene. Other specialty polymers are fiber-forming polyesters and nylon. Products considered to be high-performance engineering thermoplastics are acetal, polycarbonate, polyphenylene oxide, polyphenylene sulfide, and polysulfone. Major producers of thermoplastics (LDPE, HDPE, PP, PS, PVC) throughout the world are:

Dow	Shell
BASF	Hoechst
ICI	Union Carbide
Monsanto	Montedison
Solvay	Huls
Du Pont	Mitsubishi
Exxon	Rhone-Poulenc

There are a wide variety of polyolefins commercially available in addition to the four major polymers. Polymerization of olefins can be generally described by:

$$RCH{=}CH_2 \longrightarrow \left[\begin{array}{c} CH{-}CH_2 \\ | \\ R \end{array} \right]_n$$

where R can be:

H	Polyethylene (PE)
CH_3	Polypropylene (PP)
phenyl	Polystyrene (PS)
pyridyl	Polyvinyl pyridine
CN	Polyacrylonitrile (PAN)
COOH	Polyacrylic acid
COOR	Polyacrylates
CL	Polyvinyl chloride (PVC)

The polymerization processes for polyolefins are:

- free-radical initiated chain polymerization
- anionic polymerization
- cationic polymerization
- organometallic initiation (i.e., Ziegler-Natta)

Free-radical initiated chain polymerization is the dominant process. Anionic polymerization is employed largely in the copolymerization of olefins. Cationic polymerization is used exclusively in butyl rubber synthesis.

As described in Chapter 1, the mechanism of free-radical chain polymerization is:

initiation
propagation
radical transfer
termination

Homogeneous-solution polymerization and heterogeneous-suspension polymerization processes are generally used in free-radical initiated chain polymerization. Homogeneous-bulk polymerization is the most economical process, although there are a number of process problems associated with this method. The bulk process is used currently for the manufacture of polyvinyl chloride, polystyrene, and polymethyl methacrylate. Solution polymerization is used for synthesizing polyethylene, polypropylene, and polystyrene. Suspension polymerization is widely used to manufacture polyvinyl chloride and polystyrene (also produced by emulsion polymerization).

Bicyclic olefins are manufactured by ring-opening polymerization. An example of this is the synthesis of polynorbornene:

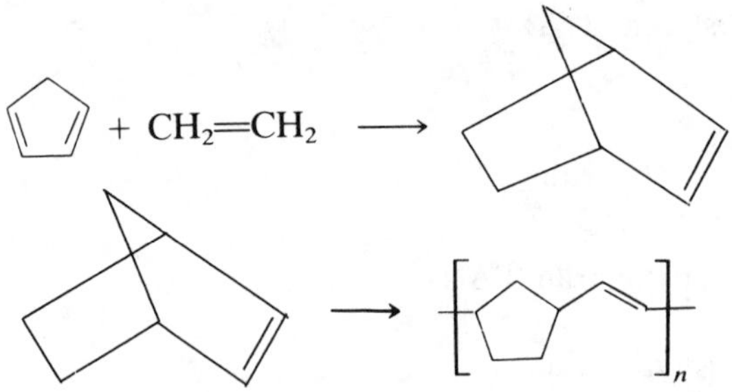

$$\text{cyclopentadiene} + CH_2{=}CH_2 \longrightarrow \text{norbornene}$$

$$\text{norbornene} \longrightarrow [\text{-cyclopentane-CH=CH-}]_n$$

The arrangement or entanglement of atoms in the chain molecules is the principle reason why polyolefins have good physical properties. Radical transfer causes branching, which has a profound effect on physical properties. Also molecular weight distribution affects physical properties as well as processability.

With substituted olefins, polymerization takes place either by head-to-tail or head-to-head polymerization; that is,

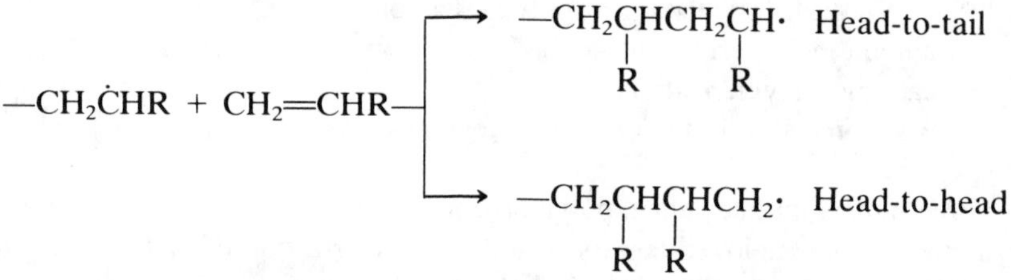

$$-CH_2\dot{C}HR + CH_2{=}CHR \longrightarrow -CH_2\underset{R}{CH}CH_2\underset{R}{CH}\cdot \quad \text{Head-to-tail}$$

$$-CH_2\dot{C}HR + CH_2{=}CHR \longrightarrow -CH_2\underset{R}{CH}\underset{R}{CH}CH_2\cdot \quad \text{Head-to-head}$$

Head-to-tail polymerization is largely favored. Head-to-head polymers are synthesized by special methods. The possible structures for substituted polyolefins are:

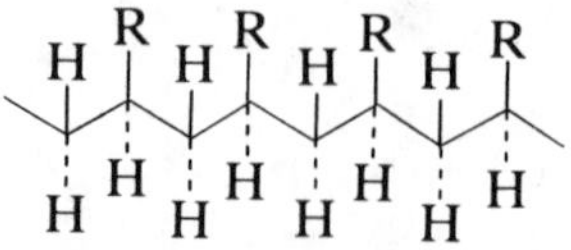

Isotactic

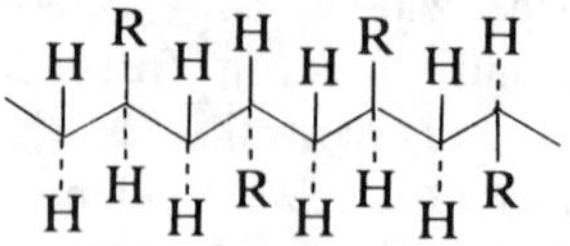

Syndiotactic

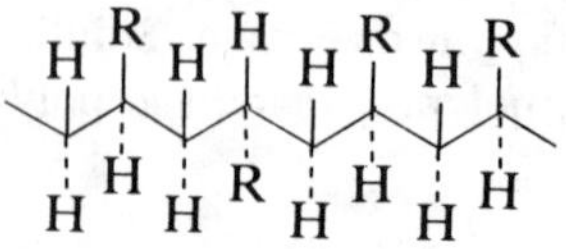

Atactic

The term *tacticity* refers to the obtainable order in the macromolecule. Atactic polymers are tandem and display the property of being amorphous or rubbery. Crystalline isotactic polypropylene has all the methyl groups above or below the horizontal plane. In the stereoregular syndiotactic polypropylene, the methyl groups alternate above and below the horizontal plane. Note that isotactic polypropylene is never truly stereoregular even with the more highly crystalline products. Degrees of isotacticity can vary from 88 percent to 97 percent.

Table 2-1 provides a summary of properties for major polyolefins.

2.2 VINYL CHLORIDE MONOMER PROPERTIES AND PVC MANUFACTURING

Vinyl chloride at room temperature and atmospheric pressure is a colorless flammable gas with a pleasant, sweet-smelling odor. The monomer boils at -13.9°C and is a colorless mobile liquid below this temperature.

The gas will burn very readily in the proper mixture of air or oxygen. The explosive limits are: lower 4.0 percent, upper 22.0 percent of volume in air. Consequently, in the production of polymers of vinyl chloride, great care must be taken so that explosions and fires do not result. Vinyl chloride in general is noncorrosive at normal atmospheric temperatures when dry. The presence of trace amounts of water and acid in monomer accelerates the formation of peroxides and polymers and the corrosion of the storage or shipping vessels. Any acetylene impurities in vinyl chloride react with copper. Properties of vinyl chloride are given in Table 2-2.

In general terms, a polymer is a large molecule built up by the repetition of small simple chemical units. In some cases, the repetition is linear, much as a chain is built up from its links. In other cases, the chains are branched or interconnected to form three-dimensional networks. The repeating unit of the polymer is usually equivalent or nearly equivalent to the monomer or starting material from which the polymer was formed. An example of this repetition is demonstrated simply by the build-up of PVC from vinyl chloride monomer.

```
H H                      H H H H H H
C=C →                   -C-C-C-C-C-C
H Cl                     H ClH ClH Cl
vinyl chloride            PVC polymer
monomer
```

The number of repeating units determines the length of the polymer chain, which in turn is referred to as the *degree of polymerization*. The molecular weight of the polymer is the product of the molecular weight of the monomer or repeating unit and the degree of polymerization.

Table 2-1 Properties of Major Polyolefins

Plastic Materials (arranged in chemical groups)			Density		Usual Appearance (solid products)				Elastic Behavior			Sample slowly heated in pyrolysis tube m = melts d = decomposes	Reaction of vapors given off al = alkaline n = neutral ac = acidic sac = strongly acidic	Ignition with small flame 0 = hardly ignitable I = burns in flame, extinguishes in absence of flame II = continues to burn after ignition III = burns vigorously, fulminates		Odor of vapors given off on heating in pyrolysis tube or after ignition and extinction	Solubility in Cold Solvents (ca. 20°C) s = soluble sw = swellable i = insoluble							
Standard Abbreviation (ISO 1043/ASTM 1600)	Name		unfilled g/cm³	filled up to g/cm³	transparent thin films	transparent, clear	hazy to opaque	usually contains fillers	leathery or rubbery, soft	flexible, resilient	hard						Gasoline	Benzene	Methylene chloride	Diethyl ether	Acetone	Ethyl acetate	Ethyl alcohol	Water
PE	Polyethylene (chlorinated PE see group 3)	soft	≧0.92		+		+			+		becomes clear, m, d, vapors are barely visible	n	II	yellow with blue center, burning droplets fall off	slight paraffin-like odor, PP and PB with different flavour	i sw	sw	i	i sw	i sw	i sw	i	i
		to hard	≦0.96							+	+						i	i sw	i sw	i	i	i	i	i
PP	Polypropylene		0.905	1.3*	+		+				+						i sw	i sw	i sw	i	i	i	i	i
PB	Polybutene-1		0.915				+			+	+						sw	i sw	i	i sw	i	i sw	i	i
PIB	Polyisobutylene		(0.93)	1.7				+	+			m, vaporizes, gases can be ignited	n	II	yellow, burns quietly	paraffin- and rubber-like	s	s	s	sw	i	i	i	i
PMP	Poly-4-methylpentene-1		0.83			+					+	m, d, vaporizes, white smoke	n	II	yellow with blue center, drips	slightly paraffin-like	sw	sw	i	i	i	sw	i	i

Table 2-2 Properties of Vinyl Chloride

Property	Value
Chemical names	Vinyl chloride, chloroethylene, chloroethene
Formula	CH_2CHC1
Molecular weight	62.501
Color	Colorless
Corrosivity	Noncorrosive when dry at atmospheric temperatures. Acetylene impurities in commercial product reacts with copper.
Odor	Faint sweet odor
Physical state	Gas at ordinary temperature and pressure. Liquid under pressure in cylinders or pressure vessels at room temperature.
Boiling point at 760 mm	-13.37°C (7.93°F)
Carbon-chlorine interatomic distance	1.69 ± .02 Angstroms with double bond character of 14%
Carbon-chlorine tetrahedral angle in VC	122° ± 2°
Charge in boiling point with pressure	
at 25°C	0.01355°C per mm Hg
at B.P. (-13.37°C)	0.03423°C per mm Hg
at t_e^a (-15.76°C)	0.03675°C per mm Hg
Compressibility factor PV/RT	
at 156.5°C	0.264
at 25°C	0.9178
at -13.37°C	0.9640
at t_e^a (-15.76°C)	0.9652
at -73.90°C (30 mm Hg)	1.0000
Critical density D_c	0.370 g/ml
Critical pressure P_c	42,000 mm Hg (55.2 atm)
Critical temperature T_c	156.6°C (313.7°F)
Dielectric constant at frequency of 10^5 and 17.2°	6.26
Dipole moment	1.44 ± .01 Debye Units[b]
Explosive limits (% by volume in air)	Lower 4%; Upper 0%
Flash point (Cleveland open cup)	-78° (-108°F)
Free energy of formation	-1890 Cal/g
Freezing point	-153.77°C (-244.82°F)
Heat of absorption	Approx. 3500 cal/mole

Table 2-2 Properties of Vinyl Chloride *(Continued)*

Property	Value	
Heat of combustion	286.2°K cal at constant pressure	
Heat of formation (estimated)	-120 cal/g (-216 Btu/lb)	
Heat of polymerization (estimated)	-272 cal/g (-490 Btu/lb)	
Ignition temperature, autogenous	472.22°C (882°F)	
Ionization potential	9.95 volts	
Latent heat of fusion	18.14 cal/g (33.12 Btu/lb)	
Latent heat of vaporization		
at 25°C	71.26 cal/g (128.27 Btu/lb)	
at 13.37°C	79.53 cal/g (143.15 Btu/lb)	
Liquid density		
at -30°C	0.99986 g/ml (8.343 lb/gal)	
at -25°C	0.99176 g/ml (8.276 lb/gal)	
at -20° C	0.98343 g/ml (8.207 lb/gal)	
at -12.96°C	019692 g/ml	
at +1.32°C	0.9443 g/ml	
at +25°C	0.9014 g/ml (7.522 lb/gal)	
at +28.11°C	0.8955 g/ml	
at +39.57°C	0.8733 g/ml	
at +48.20°C	0.8555 g/ml	
at +59.91°C	0.8310 g/ml	
Liquid density change with temperature between -30°C and -20°C	Approximately 0.00164 g/ml/°C	
Parachor	13.69	
	Frequency (cm^{-1})	**Wavelength**
Raman spectra strong peaks at	402	24.88
	721	13.87
	1,614	6.192
	3,036	3.293
	3,134	3.190
Refractive Index D line at 15°C	1.398 (calculated)	
Solubility of water in vinyl chloride	0.11%	
Solubility of vinyl chloride in water at 1 atm	0.5%	
Solubility of vinyl chloride in other solvents	Double in carbon tetrachloride, ether, ethanol, and most organic solvents	

Table 2-2 Properties of Vinyl Chloride *(Continued)*

	Frequency (cm^{-1})	Wavelength
Specific gravity	0.9121 @ 20/20°C (water = 1.00)	
Specific heat of liquid at 25°C	0.38 cal/g/°C	
Specific heat of vapor		
Cp at 25°C	0.207 cal/g/°C	
Specific heat of vapor		
Cv at 25°C	0.174 cal/g/°C	
heat capacity ratio = *Cp/Cv*	1.183	
Surface tension at -10°C	20.88 dyne/cm	
at -20°C	22.27 dyne/cm	
at -30° C	23.87 dyne/cm	
Vapor density	2.15 (air = 1.00)	
Vapor pressure		
at + 25°C	2,660 mm hg	
at -13.37°C	760 mm hg	
at t_e^* -15.76°C	692.3 mm hg	
at -55.8°C	100 mm hg	
at -73.9°C	30 mm hg	
at -87.5°C	10 mm hg	
at -109.4°C	1 mm hg	
Viscosity, liquid		
at +25°C	0.193 centipoise	
-10°C	0.248 centipoise	
-20°C	0.274 centipoise	
at -30°C	0.303 centipoise	
at -40°C	0.340 centipoise	
Volumetric shrinkage upon polymerization, approximately 35%		

[a]Temperature at which a mole of vapor in equilibrium with the liquid occupies 22.414 liters (-15.76°C).

[b]Debye Unit equal 10^{-18}, which is the product of the unit electrostatic charge of 10^{-10} e.s.u. X the order of magnitude of an atomic diameter 10^{08} cm.

Note: All heat terms are defined according to the Lewis and Randall nomenclature. A negative value indicates that heat is evolved.

Most polymers useful for plastics, elastomers, or fibers have molecular weights from 10,000 to 1,000,000. Since the molecular weight or chain length of the polymers is determined by a series of purely random events in the polymerization reaction, there is always a distribution or range of molecular weights in any finite sample of polymer. The average molecular weight may be controlled by various means, such as temperature and recipe.

Polymers can be divided into two types: condensation polymers and addition polymers. These two major classes are then further subclassified according to the raw materials from which they are derived or according to properties, chemical, process, uses, and so on.

PVC is not a member of the condensation classification of polymers, however, a short explanation of this type of reaction is given for purposes of comparison and completeness.

Condensation or step-reaction polymerization is entirely analogous to condensation of low molecular weight compounds. In polymer formation, however, the condensation takes place between two or more polyfunctional molecules to produce one large molecule with the possible elimination of a small molecule such as water, hydrogen chloride, and so on. For example, the formation of the polyester poly (ethylene adipate) is represented as follows:

$$X(HOCH_2CH_2OH) + X\left[HO\overset{O}{\overset{\|}{C}}(CH_2)_3\overset{O}{\overset{\|}{C}}OH\right] \longrightarrow H\left[OCH_2CH_2O\overset{O}{\overset{\|}{C}}(CH_2)_4\overset{O}{\overset{\|}{C}}\right]_x OH + H_2O$$

The reaction continues until all of one of the reagents is consumed.

Other polymers formed by the condensation reaction are the polyamides, polyurethanes, polysulfides, phenol-formaldehyde resins, silicones, and so on.

Many of the polymers obtained from the condensation reactions belong to a group of resins known as thermosetting resins. These resins do not soften appreciably with the application of heat and are not soluble in most solvents because of extensive three-dimensional chemical structure involved in their formation.

Polyvinyl chloride is one of the many types of resins which is formed by the process called addition or chain-reaction polymerization. This type of reaction involves reactive molecules called free radicals which contain an unpaired electron. The free radical may be formed in several ways, which is covered in a later portion of the text. One of these methods is described now in order to explain its role in addition or chain-reaction polymerization.

A free radical is formed by the decomposition (or bond cleavage) of a relatively unstable material called an initiator or catalyst.

$$C_6H_5-\overset{O}{\overset{\|}{C}}-O-O-\overset{O}{\overset{\|}{C}}-C_6H_5 \longrightarrow 2\,C_6H_5-\overset{O}{\overset{\|}{C}}-O\cdot$$

initiator or catalyst (Benzoyl Peroxide) — free radical

The free radical is capable of activating the double bond of a vinyl monomer, thereby causing it to polymerize to a polymer chain with an unpaired electron at the end of the chain.

$$C_6H_{11}-\overset{O}{\overset{\|}{C}}-O\cdot + CH_2{=}CHCl \longrightarrow C_6H_{11}-\overset{O}{\overset{\|}{C}}-O-CH_2-CHCl\cdot$$

free adical — Vinyl chloride monomer — growing or active polymer chain

In a very short time (usually a few seconds or less), many more monomer units add successively to the growing chain. Finally, two free radicals react to annihilate each other's growth activity and form a completed, now-inactive polymer molecule. Another method of ending chain growth is to add or incorporate into the recipe materials which will react with the free radicals and act as chain terminators. Controlled addition and selection of these chain-ending materials can be used to control molecular weight, the rate of reaction, or totally stop the polymerization.

As explained previously, PVC as well as other chain-reaction polymers are formed by a free-radical type of reaction.

Several different ways of forming free radicals may be used. These systems or ways of forming free radicals may be classified under four general headings:

1. Thermal cleavage of covalent bonds
2. Oxidation-reduction processes
3. Photochemical cleavage
4. High-energy cleavage

The first two systems just listed are presently the most commonly used in the manufacture of PVC.

a. **Thermal Cleavage of Covalent Bonds**

The covalent bonds in a number of organic and inorganic molecules are weak enough so that they dissociate more or less rapidly into free radicals at or near room temperature. The most useful sources of free radicals are the organic and inorganic peroxides. Some of the most commonly employed initiator or catalyst materials are as follows:

Name	Formula
benzoyl peroxide	$(C_6H_5CO)_2O_2$
lauroyl peroxide	$(C_{11}H_{23}CO)_2O_2$
caprylyl peroxide	$[CH_3(CH_2)_8CO]_2O_2$
hydrogen peroxide	H_2O_2

isopropyl peroxydicarbonate	$[(CH_3)_2CHOC(=O)]_2O_2$
potassium persulfate	$K_2S_2O_8$
azobisisobutyronitrile	$[(CH_3)_2CCN]_2N_2$

The selection of the initiator material with respect to activation, solubility, residues, and other features is highly important in the commercial production of PVC.

b. **Oxidation-Reduction (Redox) Process**

Numerous oxidation-reduction processes occur by one electron transfer step.
Such reactions are varied and some may be quite useful for free-radical polymerization. An example is the reaction of an organic peroxide with ferrous ion:

$$Fe^{++} + ROOH \rightarrow Fe^{+++} + OH^- + R\text{-}O\cdot$$

c. **Photochemical Cleavage**

Ultraviolet light may be used to activate the bond structure of vinyl chloride and other polymerizable materials producing free radicals. At this time no commercial use of this technique for production of PVC is known.

d. **High-Energy Formation**

Other ways for free-radical formation from rupture of the bond structure can be obtained by exposure to high-energy particles such as gamma rays or by mechanical means such as violent stirring or ultrasonic vibration.

Suspension polymerization is also called pearl, bead, or granular polymerization. These terms describe the polymer form and may help in an understanding of this important process of polymerization. For the purposes of this section, suspension polymerization as applied to general-purpose (calender and extrusion grade) grades of PVC and not dispersion grades will be considered. In suspension polymerization, a monomer or mixture of monomers is dispersed by strong mechanical agitation in droplet form in a second liquid phase. The monomer and polymer are essentially insoluble in this second or aqueous phase. The monomer droplets, which are much larger than those in emulsion polymerization, undergo a reaction during which time continuous agitation is maintained. To the suspending liquid which is invariably water, agents are added which function in part to prevent the coalescence of the monomer droplets during polymerization.

Polymerization initiators or catalysts soluble in the monomer phase are used to promote the polymerization reaction. This is a distinguishing feature or characteristic of suspension polymerization.

In certain cases, other agents are added to control pH, polymer porosity, or other essential properties. Furthermore, the entire reaction mass is maintained at uniform and controlled temperature. Agitation type and intensity are essential to obtain the specific polymer properties desired.

The general-purpose suspension polymers are composed of relatively large particles. The granules obtained are usually in the U.S. Standard sieve range of 40 mesh (420 microns) or 200 mesh (74 microns) or about like fine granulated sugar. It is possible, however, to employ the suspension process to produce very fine (ca 1-2 microns) particle-size dispersion-type resin. Another feature of general-purpose calendar and extrusion-grade resins made by suspension polymerization is that the polymer particles will separate or settle from the water phase if allowed to stand without agitation. In production, the water slurry of this type of suspension polymers is filtered or centrifuged to remove most of the water before drying.

The aqueous phase is continuous and serves to hold the dispersed, discontinuous monomer in the form of droplets and also as a heat exchange medium. It is the vehicle for monomer and polymer. The polymerization of vinyl chloride is a highly exothermic reaction which has been determined to be 490 Btu per pound. This heat of reaction must be safely and uniformly removed from the polymerization area. It is one of the major functions of the water to transfer this heat efficiently to the internal walls of the jacketed vessel.

The water also acts as the carrier for the suspending agents and other chemical additives in the recipe which are used to control particle size and shape and to impart other specific properties to the polymer (that is, porosity, oil absorption, and so on).

In *suspension polymerization*, the droplets of vinyl chloride monomer during polymerization reaction have a tendency to stick to each other or agglomerate. This can lead to serious difficulties from heat build-up (poor heat stability and uncontrolled molecular weight distribution) and formation of large polymer masses (usually such resin must be discarded because it is too coarse or high in "gel" content to process satisfactorily). Various additives to the aqueous phase protect the droplets and prevent undesirable agglomeration, making the process dependable. These agents are called suspending agents, protective colloids, or suspension stabilizers. Some of these agents occur naturally like gelatin, and some are synthetics like CMC, methyl cellulose, and so on. The choice of suspending agent together with the amount used is closely associated with the final resin properties such as particle size, size distribution as well as oil absorption, heat stability, color, and so on. In some cases inorganic salts, buffers, or surface-active agents are also used to impart special properties to the resins.

In suspension polymerization the monomer phase contains the dissolved polymerization catalyst or initiator. The catalysts used are materials such as lauroyl peroxide, benzoyl peroxide, caprylyl peroxide, or diisopropyl-peroxydicarbonate (IPP). With the catalyst dissolved in the monomer phase, each bead of monomer can be visualized as an individual polymerization system which is kept separated and distinct by the protective water solution of suspending agent around it. In other words the droplet is a tiny bulk or mass polymerization system.

In commercial suspension polymerizations, the ratio of the continuous water phase to dispersed monomer phase usually ranges from 1.5 to 1 to 4 to 1 by weight. This is determined by the particular product being produced. Economics direct that the volume of water be kept as low as possible, whereas quality considerations direct that enough water be present to control the heat of reaction, keep the beads from agglomerating, and so on. The temperature of polymerization is the major method of controlling molecular weight. Therefore, the specification for specific viscosity, intrinsic viscosity, or other average molecular weight measurement determines the polymerization temperature which must be used.

Emulsion Polymerization - One of the oldest methods of polymerization is emulsion polymerization. In this system the monomer is dispersed in an aqueous phase containing a surface-active agent commonly referred to as soap. These soaps or surfactants are employed in sufficient quantities to promote formation of vinyl chloride droplets much finer than is normally encountered in standard suspension polymerization. As an estimate these monomer droplets are less than one tenth or one twentieth the size of standard suspension polymer droplets. This degree of dispersion provides an unusually large surface area, causes the particles to remain dispersed under ordinary conditions, and imparts a milky appearance typical of an emulsion to the water-monomer mixture.

Almost always a soap is used to promote the initial dispersion, to make an interface layer, and to act as an emulsion stabilizer. A water-soluble initiator, or catalyst, must be used. Typical of these is potassium or ammonium persulfate. These initiators break down in the water phase and yield free radicals which migrate to the monomer-water interface and provide the monomer with the free radical necessary to initiate or continue polymerization.

In this process, the monomer is dispersed initially in droplets. Minute quantities dissolve in the water phase. This dissolved monomer is now in intimate contact with the water-soluble initiator. The water-soluble free radicals promote polymerization, producing PVC polymer. This polymer is so exceedingly fine it may form a latex, as is usually the case. In other less customary instances, conditions may be maintained so that as the polymer is formed it precipitates and mechanically pulls out monomer with it (which because it is not a solvent for the polymer remains absorbed on the polymer surface). However, the normal situation is for the polymer particles to remain in the submicron size and in a latex form. Polymerization may continue to promote the growth of the polymer particles through the solution of the adhering free monomer on the PVC-soap interface.

The soap or surfactant must perform efficiently several duties in latex emulsion polymerization. Initially it must reduce interfacial tension so that the monomer is effectively dispersed. Through promoting large surface area, it helps solubilize the monomer in the aqueous phase. It must function as a protective colloid to prevent aggregation of the growing colloidal polymer particles, and finally it must act as a suspending agent for the PVC particles in the latex.

Surfactants vary from true soaps, such as sodium or ammonium stearate, to synthetics such as alkyl aryl sodium sulfonates or alkyl sodium sulfates. Anionic emulsifiers, such as sodium alkyl sulfates or sulfonates, comprise the largest group used. Usually the alkyl group has about 12 carbons or more. Nonionic surfactants, polyethylene glycol esters, or half esters are used especially to stabilize emulsions against acids, bases, salts, or freezing temperatures. Water-soluble initiators, such as hydrogen peroxide, potassium or ammonium persulfate, sodium chlorate, and so on, are generally used. Reducing agents are often added to increase the rates of polymerization.

Solution Polymerization of Vinyls - High-quality vinyl chloride-vinyl acetate commercial copolymers are made generally using solution polymerization. A solvent is chosen which dissolves the monomers but from which the polymer precipitates when it reaches a certain molecular weight. Although precipitated, the particle remains swollen and permeable to monomer solutions and the molecular weight of the polymer continues to increase as it does in emulsion or suspension polymerization. Since few solvents or dilments are without chain-transfer effect, the molecular weight may be influenced according to the solvents selected.

Polymers obtained from solution polymerization are usually comparatively free from catalyst residues, contaminants producing haze, and very low molecular weight fractions. Quite often a solvent is chosen which can be used in the solution of polymer to be commercialized and, therefore, need not be removed from the resin. This is the case with vinyl chlorideacrylonitrile copolymers.

A major drawback to solution polymerization is the cost of the solvent and the necessity for a solvent recovery system. The system is capable of producing copolymers of outstanding clarity. Solution polymerization may be considered as a special case of suspension polymerization.

Bulk or Mass Polymerization - Bulk or mass polymerization consists in the conversion of monomer to polymer without solvents or other dispersing media. The polymerization reaction may be promoted by heat, catalysts, and/or other free-radical producing mechanisms. One advantage of bulk polymerization is purported to be that a high-purity polymer can be obtained since no solvents, water, emulsifying agents, and so on, are required. In addition, after polymerization, no aqueous drying operation is necessary. Certain economies can result from this circumstance, as well.

If this process does indeed possess disadvantages, one might expect to find them in the limited versatility of the control of the particle size and the particle-size distribution, especially at the finer side of the range. It is also not a practical process for producing copolymers. If fine particle-size resin is difficult to produce, it obviously precludes the use of the method for the preparation of dispersion-grade resin.

2.3 PVC CROSS-LINKING

This section describes how the vinyl monomer molecules join together like links of a chain to form the polymer molecule. According to generally accepted theory, PVC resins are chain-like structures of head-to-tail configurations with some branching.

$$\left(\begin{array}{cccccc} H & H & H & H & H & H \\ | & | & | & | & | & | \\ -C- & C- & C- & C- & C- & C- \\ | & | & | & | & | & | \\ H & Cl & H & Cl & H & Cl \end{array} \right)_n$$

This *does not* mean that the polymer chains are assumed to be arranged in a perfectly straight line but they are curled and/or laid out in nonuniform worm-like patterns.

In the polymerization of polyfunctional monomers such as butadiene, isoprene, divinyl benzene, and so on, where there is more than one olefinic member to the monomer molecule, the chain may grow in different directions or may tie into other polymer chains.

When the polymer particle grows in several directions, it is called *branching* or *branched-chain polymerization*. Standard polymerization of styrene-butadiene rubber is an example of this (refer to Figure 2-1.)

When the chains are chemically tied together in a net-like pattern, this is called cross-linking. Cross-linking may be accomplished in polymerization or in an after-treatment process analogous to the vulcanization of rubber. (Refer to Figure 2-2).

When cross-linking occurs in an after-treatment process, it can be called *curing*. Usually in curing, some chemical reaction or bonding takes place on the polymer. This may involve sulfur or peroxides which react with the chemical groups on the chain to form the cross-links or even simple hydrogen bonding. The cross-linking of PVC polymers can lead to increased rigidity, toughness, strength, chemical and solvent resistance, lower elongation, and so on. These properties or changes in properties due to cross-linking or cure are not always advantageous; however, some or all of these are usually desirable. Since the vinyl chloride monomer molecule has only one unsaturated linkage and the chlorine is very tightly attached to the chain, cross-linking of PVC has not been easy or commercially attractive.

Cross-linking via radiation has been accomplished and cross-linking through the use of copolymers is certainly quite a practical approach.

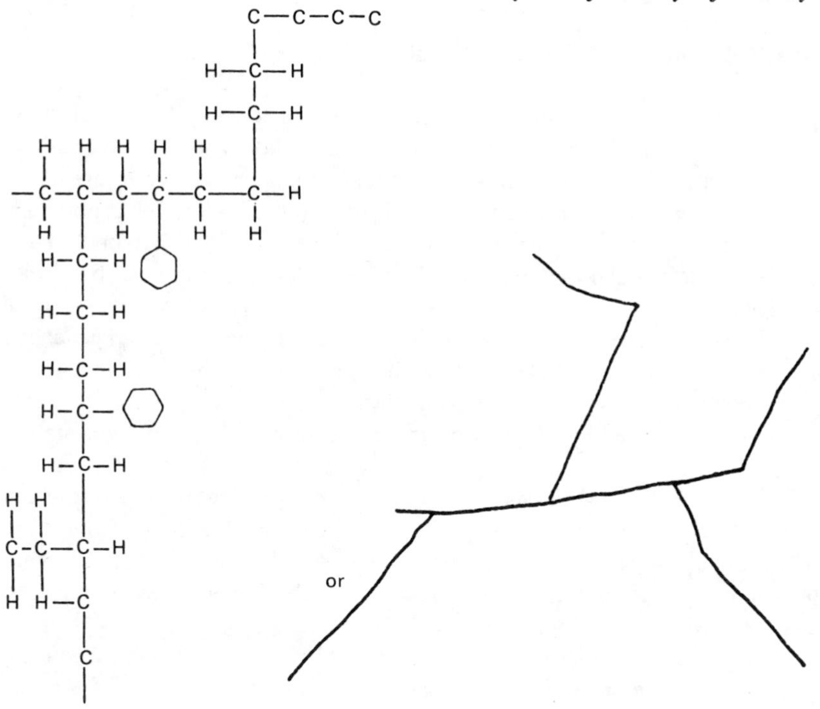

Figure 2-1 Polymerization of styrene-butadiene rubber

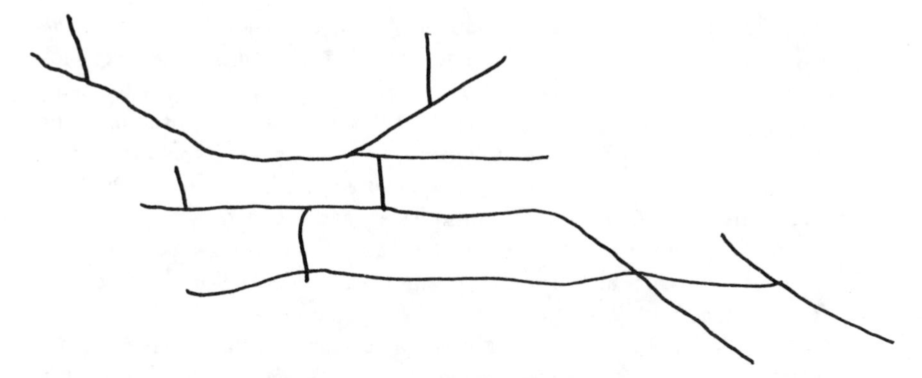

Figure 2-2 Illustrates cross-linking

2.4 COPOLYMERS AND CRYSTALLINITY OF PVC

Early on it was discovered that mixtures of monomers yielded polymers with properties that differed in an interesting fashion from homopolymers made from the predominating monomer. The individual components entered the chain molecule and remained permanently combined. Properties of the copolymers differed from those of the physical mixtures of the individual homopolymers. Thus, vinyl acetate added to vinyl chloride increases the solubility and flexibility of the molecule.

Early work emphasized the preparation and development of useful products. By the 1930s it was found that monomers differed markedly in their tendency to enter copolymers. This action differed from their ability to homopolymerize. Most of the copolymers varied in composition as formed throughout the polymerization cycle.

Empirical determinations showed other monomers varied in the rate at which they entered into copolymerization with vinyl chloride. Addition of the more active monomer throughout the batch cycle was found to reduce the heterogeneity of the copolymer. Monomers which did not homopolymerize were found to copolymerize with certain monomers.

The relative reactivities of the two monomers toward the two radicals during polymerization would determine the composition of the copolymers. These ratios are:

$$r_1 = \frac{k_{11}}{k_{12}} \quad \text{and} \quad r_2 = \frac{k_{22}}{k_{21}}$$

These reactivity ratios indicate the tendency of each monomer to attach to the growing chain. If r_1 is greater than 1 it means that M_1 radicals prefer to pick up M_1 monomers, and if less than 1 the same radical prefers M_2. It is the same case for r_2. Reactivity ratios are independent of the rate of overall reaction or the tendency of each monomer to homopolymerize. For example, styrene and vinyl acetate polymerize by themselves at about equal rates. In a copolymer of the two, the styrene adds on to the chain fifty times faster than the acetate.

An *ideal* copolymer results when the two radicals show the same preference for one monomer versus the other. Composition would depend on the reactivity ratios (if other than unity) and the monomer composition.

Alternating copolymers result if each radical prefers the other monomer exclusively. Most cases of polymerization lie between the two preceding conditions.

Most copolymers have the monomers randomly scattered along the chain and vary in composition as determined by the relative reactivities and changing monomer concentrations during the batch. By special methods using a feed of the more active monomer, it is possible to make chemically uniform copolymers. If a certain monomer ratio charge forms a copolymer of known composition, and then is fed a monomer ratio composition similar to the polymer composition, the resultant monomer-ratio concentration in the charge will remain constant and the same composition polymer will be formed. For example, if a 50/50 *A* and *B* monomer ratio makes a 70/30 *AB* initial polymer, then a 70/30 *A* and *B* monomer feed will give a constant monomer charge of 50/50 concentration which, as at the start, results in formation of 70/30 copolymer.

Other methods of copolymerization are used to make block and graft copolymers. Block copolymers are those made up where the units of monomers *A* and *B* are in long segments. These are difficult to prepare in vinyl polymerization.

Graft copolymers are special cases of block copolymers. Usually a polymer spine is prepared and a second polymer is grafted onto the spine just as is done with rose bushes or fruit trees. This is done by sensitizing the spine by catalyst, ionization, or radiation, and so on and polymerizing a second monomer with it so that this grows from the connecting link. Not all monomers capable of producing copolymers will yield block or graft copolymers.

Copolymers in general are made to modify the properties of PVC. Vinyl acetate will reduce the hardness and processing temperature as will vinylidene chloride.

Acrylonitrile is used with vinyl chloride in some commercial synthetic fiber filaments. Acrylates have also been suggested for such use.

Vinyl acetate is the most common comonomer used with vinyl chloride. Large quantities of this copolymer are used in rigid floor tile where the toughness and other physical properties such as abrasion resistance combined with ease of processing are advantageous. Synthetic lacquers also use large amounts of copolymer because of the need for easy solubility. These products use the copolymer generally with no added plasticizer. Additional flexibility in floor tile is achieved with external plasticizers as desired.

Some of the heteropolymers may have even a third monomer added to provide special properties. Enhanced adhesion to metallic substrates can result from the inclusion of carboxylic acids in the polymer molecule, as the free pendant acids can combine with (or perhaps etch) metals, and so on to be coated. Cross-linking (to increase solvent resistance) by chemical action can result from the inclusion of the proper third monomer.

In general, vinyl acetate copolymers are characterized by excellent toughness, flexibility, ease of processing, ease of application from solution, good heat sealing, and no embrittlement on aging from loss of the plasticizer.

Another major comonomer used with vinyl chloride is vinylidene chloride. This is similar chemically to vinyl chloride, having two chlorine atoms attached to the same carbon. As this is a more balanced molecule the properties are different from vinyl chloride, leading to much softer copolymers with greater solubilities than PVC. A great portion of the copolymer made has a major portion of vinylidene chloride. This makes a very tough thin film easily processed without a plasticizer. The low softening point of poly (vinylidene chloride) is advantageous for processing, and the vinyl chloride portion of the molecule greatly increases the toughness and durability. Oriented and nonoriented film produced from such polymer has many applications in the food and nonfood packaging areas. The oriented film can of course be used wherever shrinkfit packing is desired.

Filaments made from these copolymers are used in screens and woven seat covers, and so on. Paper coating and semirigid pipe are other large application areas.

Other monomers such as maleic esters are used to lend special properties to copolymers. Minor quantities of comonomer are used to give ease of processing or adhesion or low-fusion temperature in dispersion. Acrylate esters are used to change solubility, to enhance stability, or in terpolymers to promote flexibility without the use of a plasticizer. Larger molecule esters such as vinyl stearate and vinyl laureate are used to add flexibility and enhance water resistance in paints, and so on.

Special polymers with three and four monomers are made to give polymers with properties similar to those of plasticized vinyls. As almost all comonomers are more expensive than plasticizers, cost is a deterrent to the use of these terpolymers and tetrapolymers for other than specialty applications.

Commercial PVC of all types is an amorphous polymer with very few centers of crystallinity which are widely dispersed. As noted earlier, Ziegler and Natta worked on stereospecific polymers. In general, more interest has been shown in crystalline PVC. The usual Ziegler catalysts used in polyolefin polymerization react with the more polar vinyl chloride, but a number of other catalysts have been used to make stereoregular PVC, organometallics, peroxides, boron alkyls, azo compounds, and so on. Crystalline PVC is a denser polymer with a higher decomposition point and a higher melting point than commercial PVC.

The solubility of crystalline PVC is different from that of the usual PVC. It forms a cloudy solution in tetrahydrofuran and in cyclohexanone, although it clears up in the latter at + 120°C.

Most of the crystalline PVC is made at low temperatures and the lower the polymerization temperature the more crystalline the polymer. Crystalline PVC is a different polymer from the usual PVC and is not compatible with it. It does not mix well with plasticizers because of its varying solubility.

Most crystalline PVC is buried in a larger amount of the usual amorphous polymer. It can be removed by its difference in solubility, but very few practical methods for preparing crystalline PVC exist.

Crystalline PVC is usually made by low-temperature polymerization, using alkyl boron or similar organometallic free-radical catalysts. It can be made in the presence of aldehydes even at the very low molecular weight of 5,000. The aldehyde forces the stereoregularity by formation of a ring structure near the polar groups.

If the amount of vinyl chloride exceeds that of the aldehyde, then amorphous polymer is formed. With aldehydes present the crystalline polymer can be made at higher temperatures but more crystalline material is made when polymerization is cold. Lower temperatures make the polymer more linear with less branching and with a maximum of head-to-tail connections. PVC is usually almost quantitatively head to tail so the increase in this form is slight. Generally the usual polymer is branched at an average of every 70 monomer units. When made at - 40°C, the PVC whether crystalline or amorphous is not branched.

Crystalline PVC is very syndiotactic with alternation along the chain. This changes in degree only as all PVC made by free radicals is syndiotactic.

2.5 SUMMARY OF OTHER IMPORTANT POLYOLEFINS

Polyisobutylene (PIB) $$\left[-CH_2-C(CH_3)_2- \right]_n$$

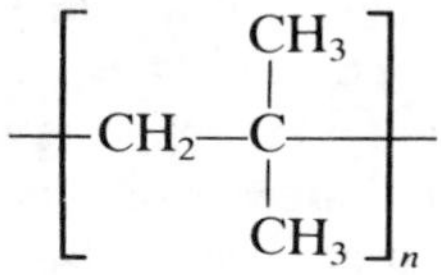

Isobutylene is polymerized continuously in methyl chloride solution at -70°C, using aluminum chloride as the catalyst. In this manner a stiff linear homopolymer is obtained. Thermoplastic PIB has a molecular weight of 1.8 million. PIB is used in lubricating oils as a viscosity-index improver, as a blending agent for polyolefins, and in the formulation of sealants and adhesives.

Polytetrafluoroethylene (PTFE) $$\left[-CF_2CF_2- \right]_n$$

Polytertrafluoroethylene (PTFE) is manufactured by the free-radical chain-polymerization process to give a high-melting homopolymer (m.p. ~327°C). The monomers are chloroform and hydrogen fluoride.

PTFE has outstanding resistance to chemical attack and is insoluble in all organic solvents. Its impact strength is high, but its tensile strength, wear resistance, and creep resistance are low in comparison to other plastics. Because of its high-melt viscosity, PTFE cannot be processed by conventional molding techniques. Molding powders are processed by press and sinter methods used in powder metallurgy. Extrusion by the ram extruder technique is also practiced. Major applications are liners and components for chemical processing equipment, high-temperature cable insulation, and molded electrical components, and reinforced PTFE applications include bushings and seals in compressor hydraulic applications, automotive applications, and pipe liners. Nonstick surfaces, applied as a coating after suitable metal preparation, for home cookware are perhaps the best-known application.

Polystyrene (PS) $$\left[-CH_2-CH(C_6H_5)- \right]_n$$

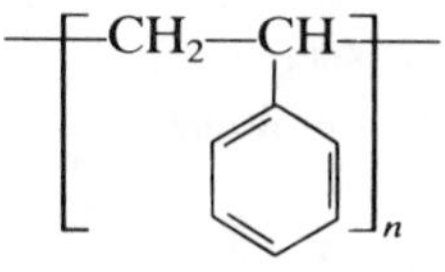

Polystyrene is made by bulk or suspension polymerization of styrene. It is commonly available in crystal, high-impact, and expandable grades. Its major characteristics include transparency, ease of coloring and processing, and low cost.

Impact polystyrene is produced commercially by dispersing small particles of butadiene rubber in styrene monomer. This is followed by mass prepolymerization of styrene and completion of the polymerization either in mass or in aqueous suspension. During prepolymerization, styrene starts to polymerize by itself, forming droplets of polystyrene with phase separation. When nearly equal phase volumes are obtained, phase inversion occurs, and the droplets of polystyrene become the continuous phase in which the rubber particles are dispersed. The impact strength increases with rubber particle size and concentration, while gloss and rigidity are decreasing. The stereochemistry of the polybutadiene has a significant influence on properties and a 36 percent *cis*-1,4-polybutadiene provides optimal properties. Uses for all types of polystyrene are packaging, housewares, toys and recreational products, electronics, appliances, furniture, and building and construction (insulation).

Polyvinyl pyridine $\left[-CH_2-CH(C_5H_4N)- \right]_n$

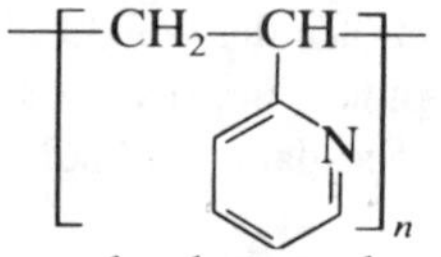

2-Vinyl pyridine can be homopolymerized by free radical or anionic polymerization. The major use for this compound is in copolymerization with other olefin monomers, especially in rubbers and fibers. In the latter the pyridine acts as a site for dye fixation. Best-known products are vinyl pyridine-butadiene-styrene terpolymers, which are composed of 70 percent butadiene, 15-21 percent styrene, and 9-15 percent vinyl pyridine. They are used as adhesives to bond textile fibers to natural and synthetic rubbers in the manufacture of tires and mechanical rubber goods. The latexes are manufactured by a batch or continuous-emulsion polymerization process. They usually contain 40.5 percent polymer solids in water.

Polybutadiene (Butadiene Rubber, BR) $\left[CH_2CH{=}CHCH_2 \right]_n$

Polybutadiene (butadiene rubber, BR) is a stereospecific (mainly 1,4-*cis*) elastomer made by solution polymerization of butadiene using Ziegler-Natta catalysts. Slight changes in catalyst composition produce drastic changes in the polymer composition.

The stereochemistry of polybutadiene is important if the product is to be used as a base polymer for further grafting. A polybutadiene with 60 percent *trans*-1,4, 20 percent *cis*-1,4, and 20 percent 1,2 configuration is used in the manufacture of ABS. The low-temperature impact strength is related to the glass-transition temperature of the components. The T of *cis*-1,4-polybutadiene is -108°C, while that of the *trans* produce is -14°C.

Butadiene rubber is usually blended with natural rubber or styrene-butadiene rubber (SBR) to improve tire tread performance, especially wear resistance. Polybutadiene is also used in the manufacture of impact-modified polystyrene products. The terms *rubber* and *elastomer* are used interchangeably, usually referring to a material that can be stretched to at least twice its original length after release of the force applied in the stretching. Relatively low-cost materials with good resiliency and durability are used in the major tire and tread market.

Polychloroprene $\left[-CH_2-\underset{Cl}{\underset{|}{C}}=CH-CH_2- \right]_n$

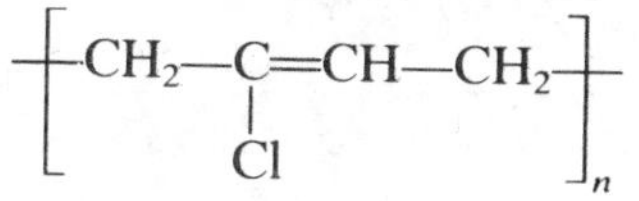

Polychloroprene was invented by the Carothers group and introduced by Du Pont in 1932 under the trade name Neoprene. Polychloroprene is manufactured by free-radical initiated emulsion polymerization, and it contains primarily linear *trans*-2-chloro-2-butenylene units arising from 1,4 addition polymerization. A total of ten stereoregular polymers are envisioned because of the asymmetry caused by the chloro group.

Compounding of Neoprene rubber parallels that of natural rubber. Vulcanization is achieved with a combination of zinc and magnesium oxide and added accelerators and antioxidants. Carbon black or mineral fillers are sometimes added also. Adhesive-grade polychloroprene often contains other compatible resins. Industrial rubber goods include uses such as conveyor belts, diaphragms, hose, seals, and gaskets. Automotive uses include hose, V-belts, and weatherstripping. Some major uses in construction are for highway joint seals, pipe gaskets, and bridge mounts and expansion joints.

Polyisoprene $\left[-CH_2-\underset{CH_3}{\underset{|}{C}}=CH-CH_2- \right]_n$

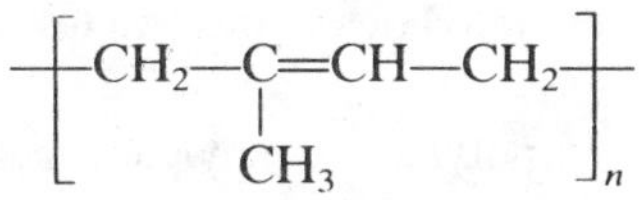

Polyisoprene is synthesized by stereospecific solvent polymerization, using Ziegler-Natta catalysts. The *cis*-1,4-polyisoprene is chemically identical to natural rubber but is cleaner, lighter in color, more uniform, and less expensive to process. Copolymers of isoprene with butadiene are also readily obtainable because the reactivities of both monomers are the same. For example, using Al(*i*-Bu)3/TiCl4 as the catalyst, both monomers enter into the copolymer chain in the same configurations (98 percent *cis* for isoprene, 70 percent *cis* for butadiene) as in the homopolymers.

Polyisoprene elastomers are used in the construction of passenger car tire carcasses and inner liners and also in the manufacture of heavy-duty truck and bus tire treads. The increasing demand for radial-ply tires in the United States is likely to increase the use of isoprenic elastomers (blends) for passenger car and truck tires. Mechanical goods, footwear, sporting goods, sealants, and caulking compounds are other important applications for polyisoprene elastomers.

3 *PROPERTIES AND NOTES ON POLYETHYLENE AND POLYPROPYLENE*

3.1 GENERAL DESCRIPTION

Polyethylene and polypropylene are major volume leaders in polyolefin polymers. This chapter contains general properties information and notes on these important plastics.

Polyethylene (PE) has the following structure: $[CH_2CH_2]_n$.

The two principle types of PE are low-density (LDPE) and high-density (HDPE). LDPE is manufactured under high pressure using peroxide initiators. Typical conditions are 15,000 to 50,000 psi and temperatures up to 350°C. The amorphous, branched LDPE has a melting range between 107° to 120°C. Typical levels of consumption are about 3,502,000 tons in the United States, 3,950,000 tons in Western Europe, and 1,360,000 tons in Japan. LDPE is made by free-radical initiated chain polymerization. Approximately 55 percent of the world's production goes into film, 10 percent into housewares and toys, and 5 percent into cable and wire coatings.

The monomer is ethylene. To polymerize ethylene, the monomer must be available in high purity. Polymerization catalysts are added to the monomer in concentrations typically less than 0.005 weight percent. The gaseous monomer is compressed before it is heated in the reactor. Under process conditions in the reactor, ethylene is an incompressible liquid. This means that the reaction takes place in ethylene solution. The unspent ethylene is recycled without any purification. The molten polymers produced are extruded and granulated.

High-density PE (HDPE) is based on Ziegler-Natta catalysts or metal oxide-catalyzed chain polymerization. The latter is a chromium oxide catalyst. This is a low-pressure process. The melting range is between 130° to 138°C. Approximate levels of consumption are 2,182,000 tons in the United States,

1,450,000 tons in Western Europe, and 750,000 tons in Japan. The approximate breakdown of market uses is: 40 percent bottles; 35 percent toys, housewares, containers; 10 percent pipe and piping components; 5 percent film and sheet.

For HDPE, the Ziegler-Natta catalysts are generally a mixture of titanium chloride and chlorodiethylaluminum with aluminum-to-titanium ratios ranging between 1:1 and 2:1. Polymerization conditions are carried out at temperatures below the polymer's melting point in hydrocarbon solvents which serve as dilutents. During the polymerization reaction, the insoluble polymers precipitate out of solution. Reaction temperatures are typically in the range of 50° to 70°C and reactor residence times are between one and four hours. In contrast, the solid-supported chromium oxide-catalyzed reaction is run continuously in a hydrocarbon solvent at temperatures of 125° to 160°C.

In general, polyethylene can be described as a partially amorphous and partially crystalline plastic. The degree of crystallinity is controlled through side-chain branching. LDPE has crystallinities as low as 50 percent. HDPE has up to 90 percent crystallinity. The properties of stiffness, tensile strength, hardness, heat and chemical resistance, and barrier properties increase with density. Impact strength and stress-crack resistance are reduced with increasing density.

Linear low-density PE (LLDPE) is produced by a low-pressure process. LLDPE polymers are linear but have a significant number of branches. This material displays high strength due to linearity and derives toughness from branching. Densities for all three types of PE range as follows:

LDPE	0.912 - 0.94
LLDPE	0.92 - 0.94
HDPE	> 0.958

PE can be transformed from a thermoplastic material into an elastomer via low levels of chlorination (typically 22-26 percent chlorine). These products are softer and tend to display elastomeric or rubbery characteristics. Additionally, they are more soluble and compatible. Crystallinity is reduced as a result of reduced chain order due to tandem substitution.

Polypropylene (PP) has the following structure: $\left[-CH_2\underset{\underset{CH_3}{|}}{C}H- \right]_n$

The monomer for PP is propylene. This material is predominately produced by low-pressure processes based on Ziegler-Natta catalysts (that is, aluminum allyls and titanium halides). The vast majority ($>$90 percent) of the polymer is in the isotactic form. Isotactic PP does not crystallize like PE (that is, planar zigzag). Steric hindrance by methyl groups prohibits this conformation. Instead, isotactic PP crystallizes in a helical form whereby there are three monomer units per turn of the helix.

PP has a wide range of applications, ranging from fiber and filaments to films and extrusion coatings. PP fibers are manufactured by an oriented extrusion process. Two important advantages of PP are its inertness to water and microorganisms and it is a low-cost polymer (around $0.27 - 0.33 per pound). Typical applications include carpet backing, upholstery fabrics, carpet yarn, and interior trim for automobiles. The market breakdown is roughly: 30 percent fibrous products; 15 percent automotive parts; 15 percent packaging products; 5 percent appliance parts; 5 percent toys and housewares. Approximate consumptions are 1,763,000 tons in the United States, 1,250,000 tons in Western Europe, and 900,000 tons in Japan.

3.2 CHEMISTRY AND PROPERTIES NOTES

The remainder of this chapter contains general information on the properties, chemistry, and processing characteristics of two major commodity plastics; polyethylene (PE) and polypropylene (PP). Information and data presented are in a condensed format for quick reference by the user.

Figure 3-1 describes the structures and conceptual processes for these polymers.

Figure 3-2 depicts the types and specific structures of polypropylene polymers. The important PP product features for commercial PP catalysts are:

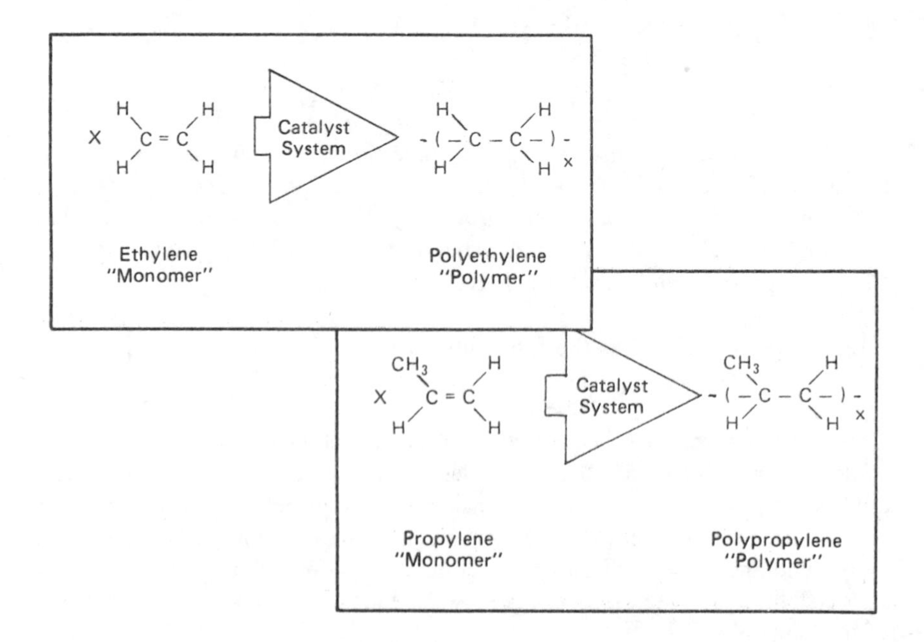

Figure 3-1 Illustrates basis for polymerization processes

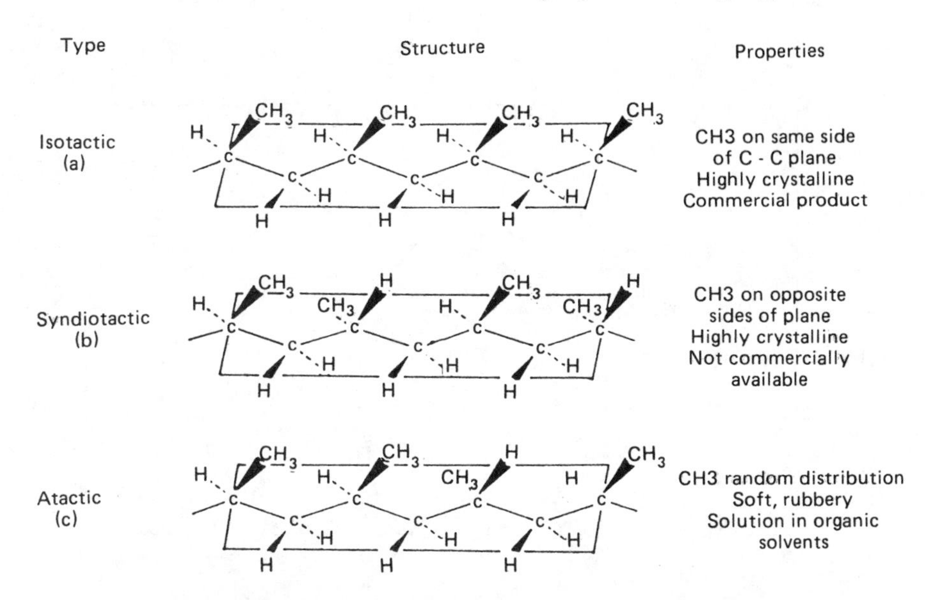

Figure 3-2 Illustrates polypropylene polymer types

- high isotacticity index (high HI)
- high catalyst productivity (that is, gms of polymer per gms of catalyst).

The aim in a commercial process is to achieve the lowest amount of catalyst residue in the polymer. Catalyst residues affect polymer stability, color, corrosivity, and general properties.

- polymer particle shape—a spherical shape is desirable for efficient polymer flowability in downstream processing operations.
- polymer particle-size distribution—a narrow particle-size distribution is desirable from the standpoint of processing. Particle-size distribution is rated by the "steepness" of the particle-size distribution curve and is quantified by the Particle-Size Distribution Index (PSDI). An example of a particle-size distribution curve is shown in Figure 3-3.
- polymer average particle size—the desired size depends on process and downstream applications.
- amount of polymer fires—fires are undesirable because they cause handling problems and fire hazards. As a rough guide, most manufacturers try to achieve 1 percent or less particle sizes under 106 microns.
- molecular weight distribution (MWD)—this is technically measured by the weight-to-number average (Mw/Mn) molecular weight ratio as measured by gel permeation chromatography (GPC). A controlled MWD is desirable for a range of polymer applications.

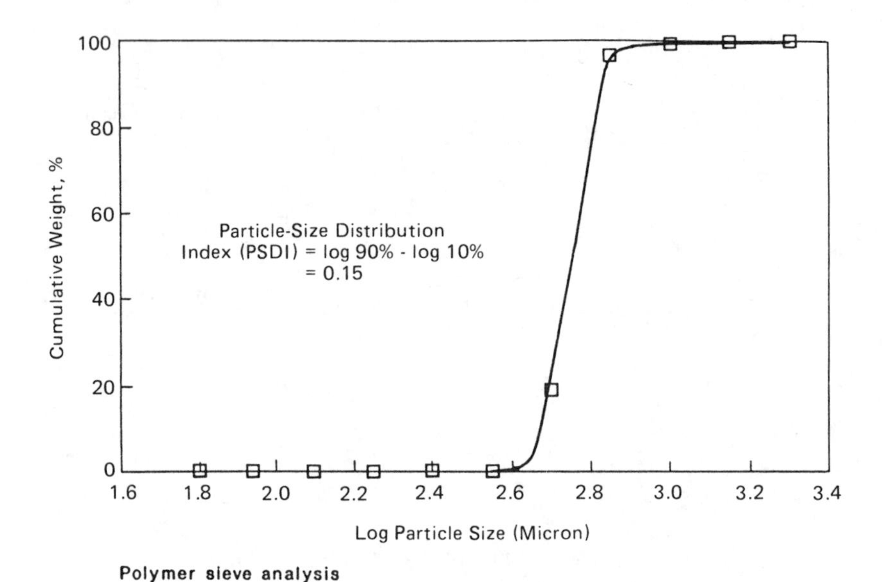

Figure 3-3 Particle-size distribution example for polypropylene

Figure 3-4 shows the breakdown of commercial PP production. Manufacturing processes produce both isotactic and atactic PP. The isotactic index is measured by the amount (percent) of heptane-insoluble material. Heptane insolubles can be controlled by the design of the catalyst system and/or by process polymer solvent extraction. Examples of catalyst systems are given in Figure 3-5.

Manufacturing Process Producrs Both Isotactic
and Atactic Polypropylene

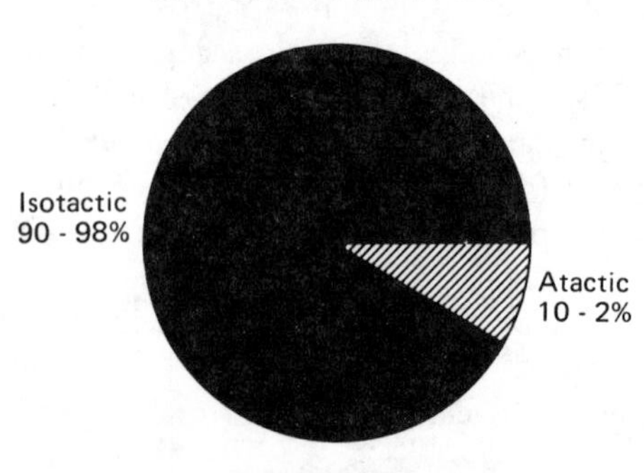

• *Isotactic Index* is Measured By The Amount (%) of *Heptane-Soluble* Material (Hi)

• Can Control Hi By Design of Catalyst System and/or By Process Polymer Solvent Extraction

Figure 3-4 Commercial polypropylene production

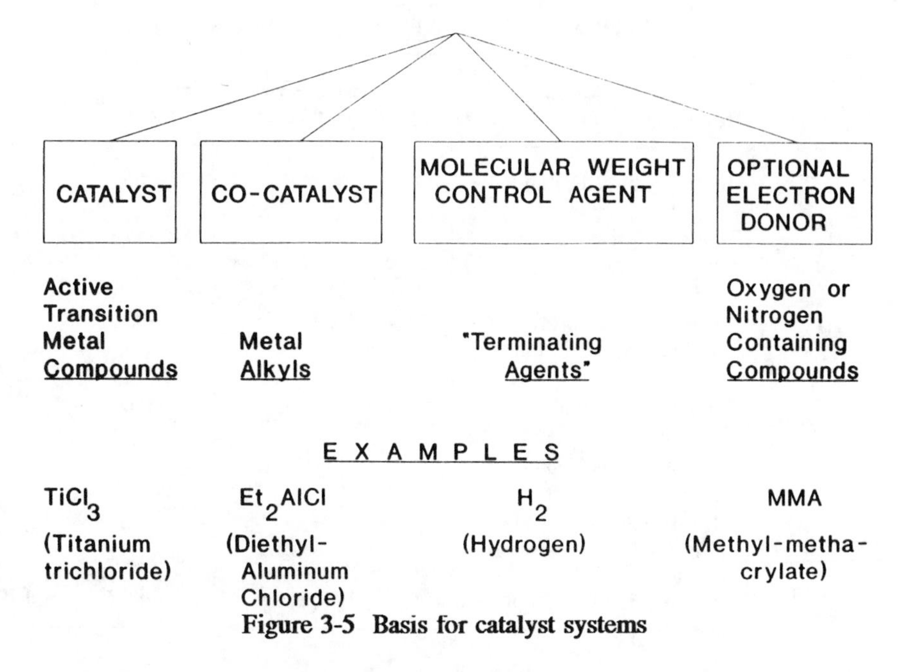

Figure 3-5 Basis for catalyst systems

Tables 3-1 and 3-2 provide a summary of PP catalyst development stages. Table 3-3 provides a summary of properties of commercial PP catalysts.

Table 3-1 PP Catalyst Development Stages

Stage	Description	Year	Catalyst System
I	Insitu produced catalyst	1957	$TiCl_4 + Et_3Al + H_2$
II	Externally produced catalyst "1st-Generation"	1960	$TiCl_3 \bullet 1/3AlCl_3 + Et_2AlCl + H_2$
III	Electron-Donor promoted "1st-Generation" Catalyst	1970	$TiCl_3 \bullet 1/3AlCl_3 + ED + Et_2AlCl + H_2$
IV	"High-surface area" "Low-aluminum $TiCl_3$: "2nd-Generation" Catalyst	1976	$TiCl_3 \bullet xAlcl_3 \bullet y^E + ED + Et_2AlCl + H_2$ (E = Ether)
V	"$MgCl_2$-supported $TiCl_4$" "3rd-Generation" Catalyst	1979	$MgCl_2 \bullet TiCl_4 \bullet ED + Et_3AlCl \bullet ED + H_2$
VI	Improved 3rd-Generation Catalyst	1980	Used better ED's better preparation.

Table 3-2 PP Catalyst Development Stages

Stage	Description	Relative Activity (Wt. Poly/Wt. Ti)	Hi (%)
I	Inisitu produced catalyst	1	75–85
II	Externally produced catalyst "1st-Generation"	10	88–91
III	Electron-Donor promoted "1st-Generation" Catalyst	12	91–93
IV	"High-surface area" "Low-aluminum $TiCl_3$: "2nd-Generation" Catalyst	40	94–96
V	"$MgCl_2$-supported $TiCl_4$" "3rd-Generation" Catalyst	1000	93–98 (Controlled)
VI	Improved 3rd-Generation Catalyst	3000	93–98 (Controlled)

Table 3-3 Properties of Commercial PP Catalysts (Development Stages)

Description	Catalyst Productivity (g Poly/g Cat)	Polymer Properties Size (u)	% Fines (<106u)	BD (#/CF)	MWD
"1st-Generation" Catalyst ($TiCl_3 \bullet 1/3AlCl_3$)	2,500	250	20	28	B
"2nd-Generation" or Low-Aluminum Catalyst ($TiCl_3 \bullet xAlCl_3 \bullet yED$)	10,000	650–1000 (Controllable)	1	32	N
"3rd-Generation" Catalyst ($MgCl_2$-Supported $TiCl_3$ + ED)	10,000	250	20	27	B
Improved 3rd-Generation Catalyst (Used better ED, and prepartion)	30,000	400–3,000 (Controllable)	0.2	30–32	NN

B=broad
N=narrow

Figure 3-6 provides an overview of the PP polymerization mechanism. Figures 3-7 and 3-8 illustrate the features of crystal structure and catalyst support.

Major product applications for PE are listed in Table 3-4 and end-use applications are given in Table 3-5.

PROPYLENE POLYMERIZATION MECHANISM
(Cossee's Monometallic Mechanism)

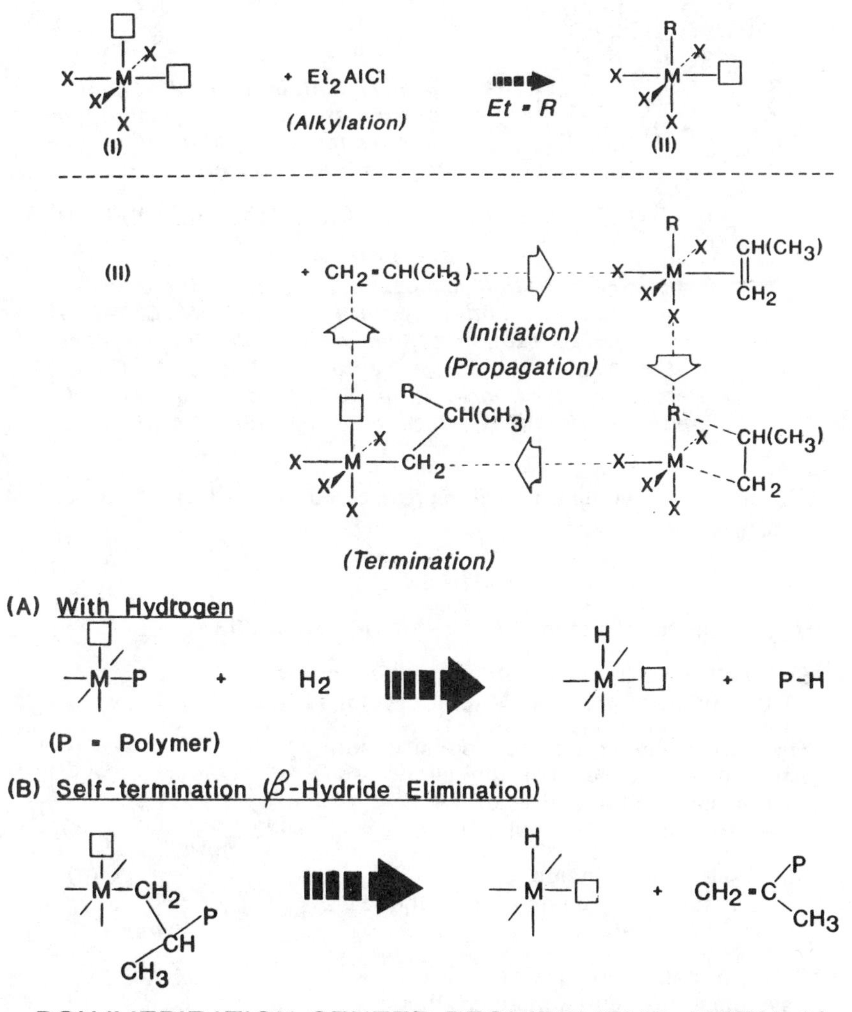

POLYMERIZATION CENTER REQUIREMENT: *METAL M WITH 6 SITES AROUND AND 2 ARE VACANT*

Figure 3-6 Illustrates propylene polymers

TiCl3.1/3AlCl3 Crystal Structure (Two-Dimensional Representation)

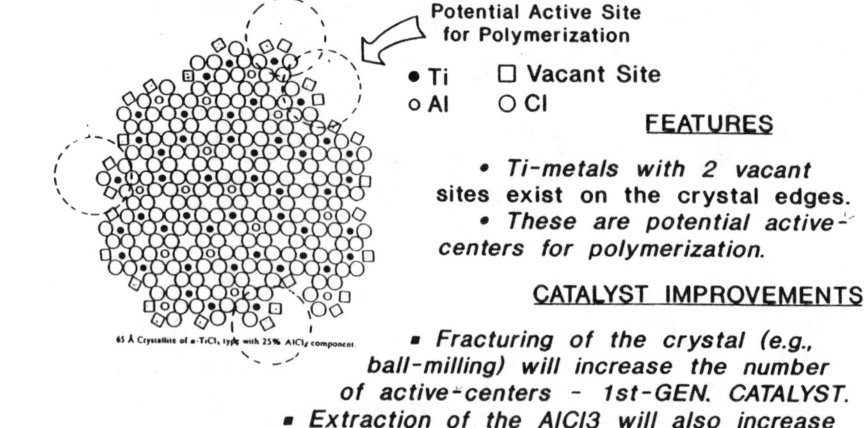

Figure 3-7 Two-dimensional representation of TiCl3.1/3AlCl[3] crystal structure

FEATURES

- MgCl2 has a structure similar to TiCl3.1/3AlCl3.
- By itself MgCl2 does not polymerize propylene, hence, TiCl4 is supported on MgCl2 crystal lattice.
- The difference in electron density and spatial environment of the supported TiCl4 necessitates use of an electron donor (internal ED).
- Also needs an outside electron donor (ED') to complex with the Et3Al co-catalyst.
- Catalyst and cocatalyst systems are represented in the figure.

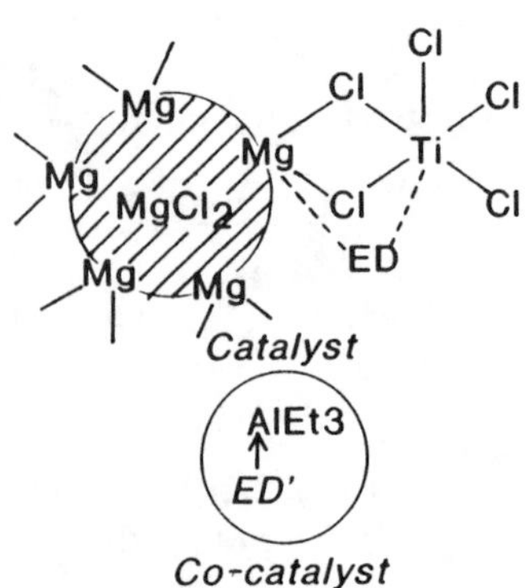

Figure 3-8 Features of Mg Cl2 as catalyst support

Table 3-4 Major Product Applications

Products	Denisity (G/CC)
HDPE (High Denisty Polyethylene)	
- Homopolymer	> 0.959
- Copolymer	0.941−0.959
MDPE (Medium Denisty PE)	
- Copolymer	0.926−0.940
LLDPE (Linear-low Denisty PE)	
- Copolymer	0.910−0.925
Broad MWD HDPE and MDPE	0.926−00.960

Table 3-5 Polyethylene End-Use Applications (Broad MWD PE)

Use	Broad MWD (MIR = 50−150)		
	Denisty	HLMI	MI
High-strength film	0.951	16	
Blow molding	0.955	20−35	
Pipe	0.953		0.10
Cable	0.951	16	

MI = I2, Melt Flow Index at 2 kg weight
HLMI = I21, Melt Flow Index at 21 kg weight
MIR = Melt Flow Index ratio, MI/HLMI

A summary of major considerations in PE catalyst development is given in Table 3-6. Table 3-7 provides an overview of catalyst systems for high-density PE (HDPE) and linear low-density PE (LLDPE).

Examples of catalysts used for HDPE and LLDPE are given in Table 3-8.

Tables 3-9 through 3-13 provide summaries of the commercial catalyst technologies used for PE. Table 3-14 lists some catalyst incompatibility issues regarding the Union Carbide Corp. (USS) process.

Table 3-15 qualitatively describes process control parameters in achieving product properties.

Table 3-16 lists reactor design bases for PE processes based on Ziegler-Natta catalysts.

Table 3-5 Polyethylene End-Use Applications (*Continued*) (Narrow MWD PE)

Use	Narrow MWD (MIR = 20−50) Denisty	MI
General-purpose film	0.918	1
	0.918	2
Injection molding	0.926	50
	0.930	40
Cast Film	0.918	2
	0.918	3
Rotational molding	0.934	5
	0.945	2.5
Injection Molding	0.953	30
	0.963	8

Table 3-6 Major Considerations in Polyethylene Catalyst Development

Process Operation	Product Properties
Costs - Manufacturing - Storage/distribution	High Productivity - For nondeashing process -Ash levels: Ti < 5 ppp Cl < 50 ppm Al < 50 ppm
Catalyst Morphology - Physical form for feeding - Gas phase versus slurry process	
Good Hydrogen-MFI Response	Polymer MW and MWD Control -Mw/Mn of 3 to 40
Good Comonomer Incorporation	Particle Morphology - Shape - Porosity - Bulk density
Operability - Reactor stability - Copoplymer "stickiness"	
Compatability between catalyst systems	Copolymer Properties - Branching (density) - Composition - Sequence Length Distance

Table 3-7 Overview of Catalyst Systems for HDPE/LLDPE

Support	Preparation Methods	Advantages/Comments
Magnesium Chloride	Ball-Milled/Ti-Cmpd.	Poor Particle Form High Chloride Level Narrow MWD PE
	Solubilized Mg-Cmpd. + Ti- Compound	Good Particle Form High Chloride Level Narrow MWD PE
Silica	Encapsulate $MgCl_2$-$TiCl_3$ Complex	Low Chloride Level Particle Form From Silicia Particle Narrow MWD PE
	Impregnate Chromium Compounds	No Chlorides Broad MWD PE
	Impregnate Vanadium Compounds	Low Chloride Level Broad MWD PE

Table 3-8 Catalyst Examples for HDPE/LLDPE (Magnesium Compound Supported)

Catalyst Preparation	Company
Ball-Milled Type $MgCl_2 + TiCl_4$ -- (Ball-milled)	Montedison
Solubilized Mg-Cmpd. -$MgCl_2 + 6EtOH + EtAlCl_2 + TiCl_4$	Mitsui Petrochemical
-$MgR_2 + 0.17AlR_3 + HCl + Ti(Opr)_4$	Dow Chemical

Table 3-17 provides a list of handling and safety issues of standard catalyst materials.

Figure 3-9 provides a process flowsheet for the gas-phase polymerization process of LLDPE. Table 3-18 provides LLDPE typical operating conditions.

Figure 3-10 provides a product property map, relating melt index to resin density.

Figures 3-11 and 3-12 provide summaries of HDPE operating modes and a process flow scheme, respectively.

Figure 3-13 provides a review of PP chemistry. A comparison of catalyst systems for PP is given in Table 3-19. Table 3-20 provides a summary of PP product attributes and uses.

Figures 3-14 through 3-18 provide process flowsheets for PP production schemes.

Table 3-9 Catalyst Examples (Silica Supported)

	Preparation Steps	Company
(A)	Calcined (dried) Silica + $MgR_2/0.17AlR_3 + TiCl_4$	CHEMPLEX
(B)	Chromium(VI) Oxide (CrO_3) + Calcined SiO_2 + Additives 600-800oC in Air Catalyst	Phillips Petroleum
(C)	Calcined Silica + Cp_2Cr	Union Carbide Corp. (UCC) HDPE Catalyst

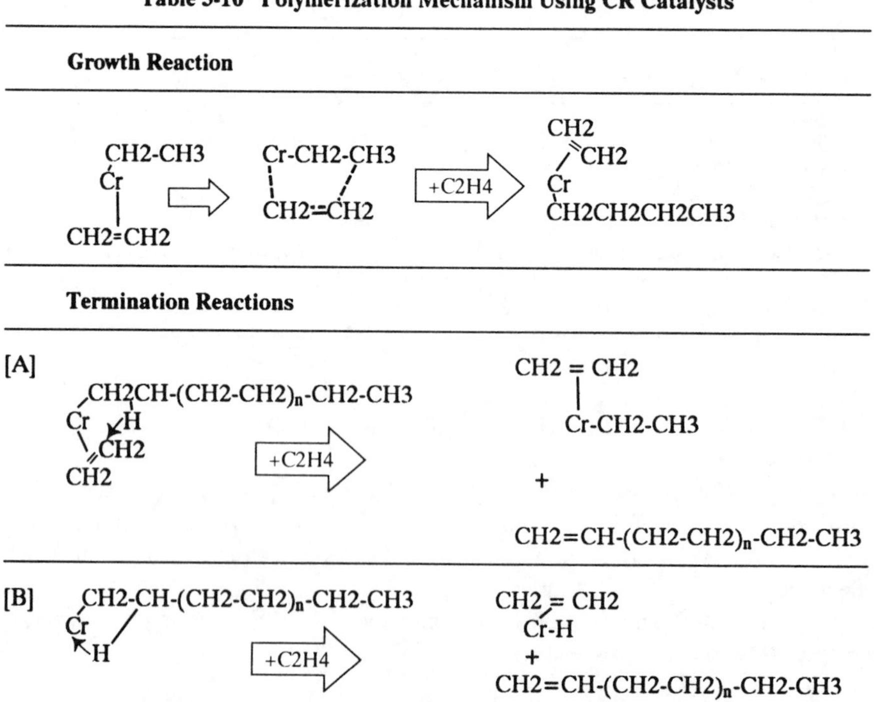

Table 3-11 Mechanistic Approach to Silica-Supported Catalysts (Chromium-Based Catalyst for Broad MWD PE)

Reaction of CrO3 with Support

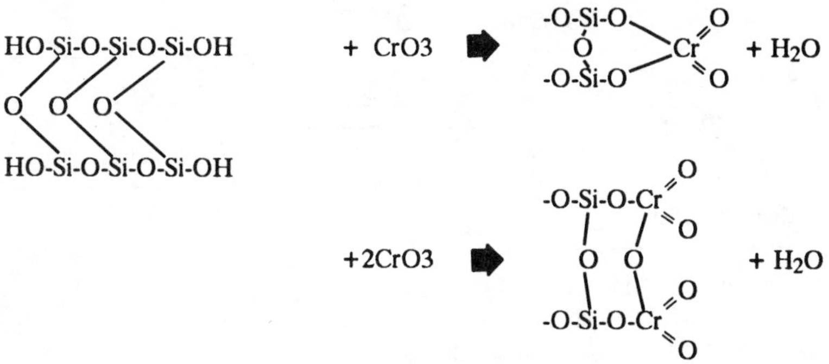

HO-Si-O-Si-O-Si-OH / O O O / HO-Si-O-Si-O-Si-OH + CrO3 ➡ -O-Si-O, O, -O-Si-O (cyclic) Cr(=O)(=O) + H_2O

+2CrO3 ➡ -O-Si-O-Cr(=O)(=O) / O O / -O-Si-O-Cr(=O)(=O) + H_2O

Reduction of Cr(VI) to Cr (II)

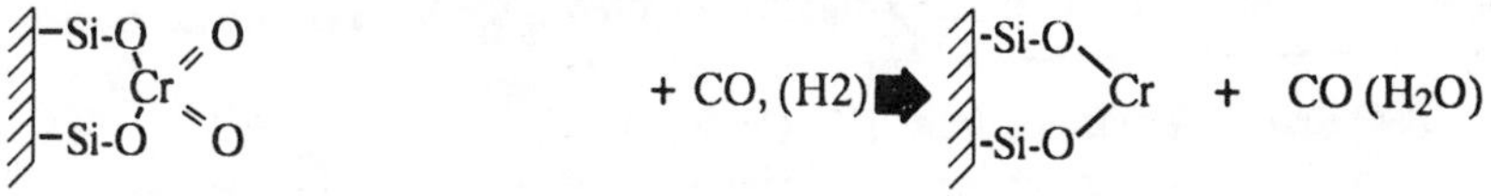

–Si-O, –Si-O Cr(=O)(=O) + CO, (H2) ➡ -Si-O, -Si-O Cr + CO (H_2O)

Table 3-12 Union Carbide Corporation Catalysts for LLDPE (UNIPOL Catalysts)

M-Catalyst Preparation (UCC Patent US 4,719,193)
Steps

A Silica Calcined at 600°C + TEAL

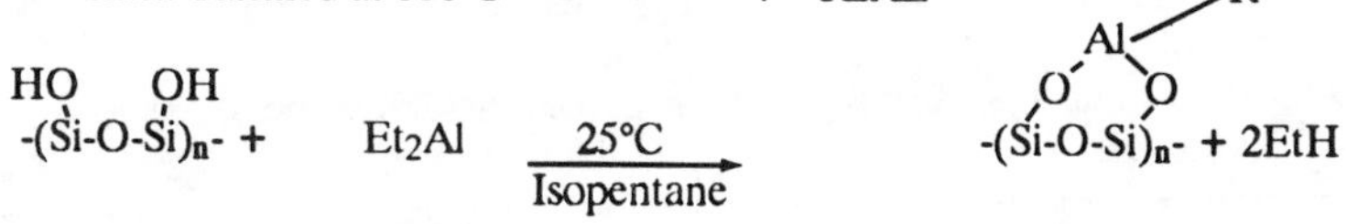

HO OH -(Si-O-Si)$_n$- + Et_2Al —25°C, Isopentane→ (O-Al(R)-O) -(Si-O-Si)$_n$- + 2EtH

B Precursor Preparation

2.5$MgCl_2$ + $TiCl_4$ —60°C, THF Solvent→ $(MgCl_2)_{2.5}$ $TiCl_4(THF)_7$

C Product A + Product B —50°C, THF→ "Silica-Impregnated Precursor"

D Pre-reduction

Product C + 1. DEAC 2. TRI-nHexyl aluminum —50°C, Isopentane→ "M-1" Catalyst

TiCl3-based M-1 Catalyst
Replace Step B with

$MgCl_2$ + $TiCl_3$ • 1/3$AlCl_3$ —60°C, THF Solvent→ [Mg_A Ti_B Al_C] Cl_n $(THF)_X$

Table 3-13 Union Carbide Corporation Catalyst for BMWD PE (UNIPOL Catalysts)

F-4 Catalyst for Molding and Extrusion Resins

Preparation (UCC Patent US 4,011,382)

$SiO_2 + CrO_3 \xrightarrow{150\text{-}200°C} SiO_2 \cdot CrO_3$

$SiO_2 \cdot CrO_3 + Ti(OR)_4 \longrightarrow$ [I]

[I] + $(NH_4)_2SiF_6 \xrightarrow{800°C}$ F-4 Catalyst

Characteristics

Mw/Mn = 13 - 15 Versus 3 - 5 for M-Type Catalysts
Ti Improved Mi Response
F Improved Comonomer Response

Table 3-14 Some Catalyst Incompatability Requires a Special Transition Procedure

UCC Reactor Transition from Ti to Cr-Based Catalysts *

Ti catalyst feed is interrupted to allow drop of reaction rate to 10-30% of initial rate (4 Hrs).

A chemical scavenger is circulated in the reactor to consume remaining aluminum alkyl cocatalyst (4 Hrs).

Cr catalyst is introduced into the reactor where polymerization usually begins within 2 hrs. Full production rate is achieved after 10 Hrs.

Ti containing product is purged from the reactor in about 4 reactor turnovers.

* UCC Patent

Table 3-15 Process Control of Product Properties

Process Parameter (Increasing)	Molecular Weight	MWD	Denisty
Temperature	↓	—	↓*
Pressure	↑	—	—
Comonomer Concentration	↓	↓*	↓
Hydrogen Concentration	↓	—	—
Cocatalyst (Aluminum Alkyl)	↓	—	—

Remark: Primary control of MWD is catalyst selection, not process control.

* Small effect.

Table 3-16 Reactor Design Bases for PE Process Using Ziegler-Natta Catalysts

	Gas-phase (UCC)	Slurry (Phillips)	Solution (DuPont)	High/Med Pressure (CDF/DOW)
Pressure (psig)	300	500	≈2,000	7,000-20,000
Temperature (°C)	80-112	90-110	220-260	200-300
H2/C2 (Molar)	0.01-0.2	0.01-0.02	—	0.004-0.02
C4/C2 (Molar)	0.01-0.4	0.01-0.3	0.01-0.3	0.01-0.3
Residence Time (hrs.)	3-5	1.5	2-5 min.	20 sec.- 2 min.
Solids conc. (%)	≈70	15-20	≈10	15-30
RX. Diluent	None	Isobutane	Cyclohexane	(Butene - 1)

3.3 NOTES ON PLASTICS FINISHING STEPS

This section contains a series of notes and diagrams pertinent to plastic-finishing operations. Figure 3-19 provides a summary of typical plastics-finishing steps.

Figures 3-20 and 3-21 illustrate direct feeding processes for resins and additives, without and with preblending.

Figure 3-22 illustrates a process flowsheet for additive addition for granular masterbatching.

Figure 3-23 illustrates conceptually additives addition in melt-fed extruder operations. The principle objectives in finishing extruder functions are

- Convert granules to pellets (melting, pumping and pressurization)
- Convert melt to pellets (pumping and pressurization)
- Incorporate additives (mixing)

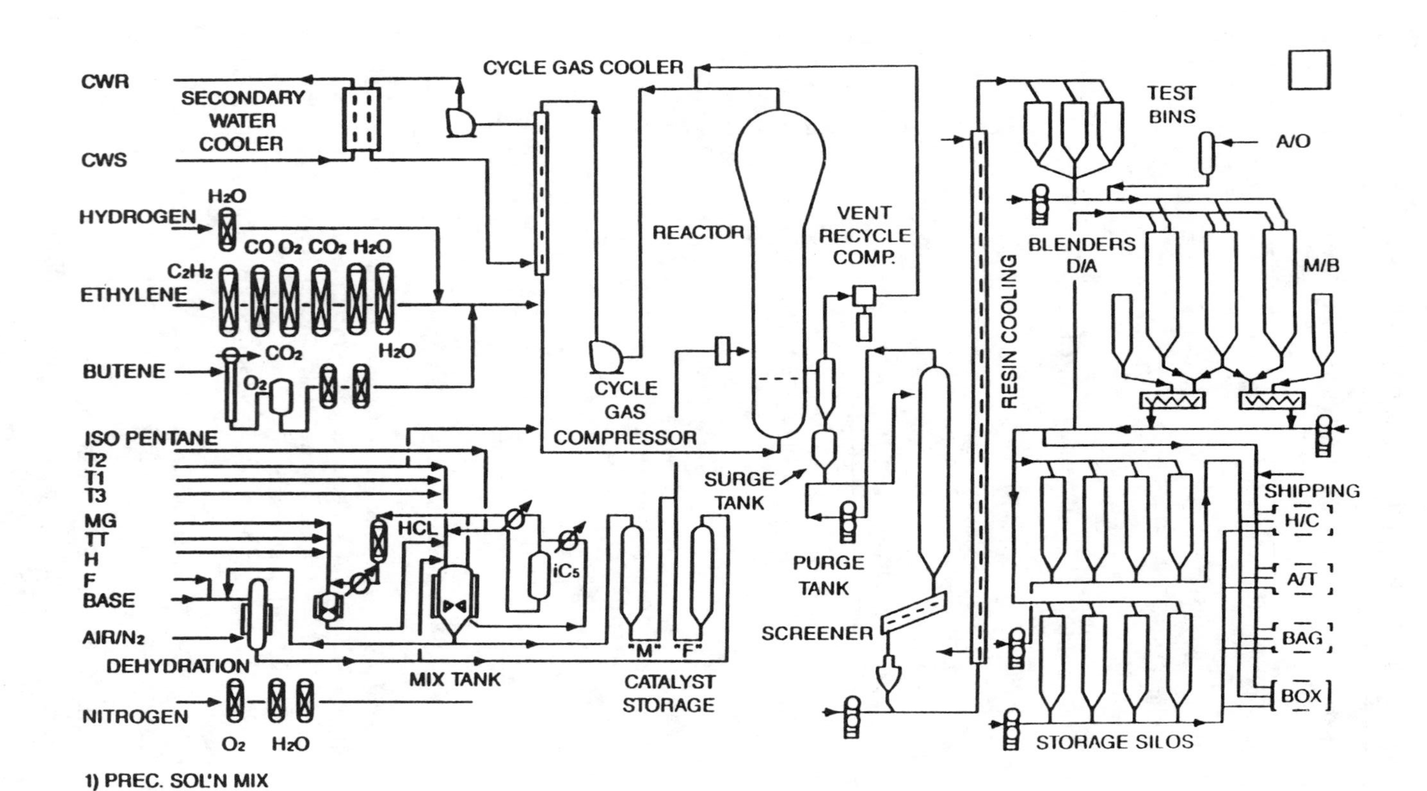

Figure 3-9 Process flow diagram for the Unipol gas phase process

Table 3-17 Handling and Safety Aspects of Some Catalyst Materials

Aluminum Alkyls

Typically clear colorless mobile liquids which are miscible in all proportions with hydrocarbons.
React violently with air or water; can be pyrophoric (flames up) when concentrated or neat.
Can be handled safely as dilute hydrocarbon solutions and in nitrogen atmosphere.
Low corrosivity - common metals are not attacked.

Magnesium Alkyls

Same handling techniques and precautions are necessary as with aluminum alkyls.
Aluminum alkyls or ethers are added in small quantities to control solution viscosities.

Titanium Tetrachloride

High boiling point, straw-colored liquid miscible with hydrocarbons.
Produces white fumes in contact with moist air to give off hydrochloric acid fumes.
Handle in an inert atmosphere.
Corrisive nature makes use of chloride-resistant steel desirable.

Silica

Typically small particle-size amorphous material which should be handled as a nuisance dust.
Contains up to 15% adsorbed moisture which needs to be removed prior to use as catalyst support.

Caution: Handle materials with care and after treating.

Table 3-18 LLDPE Typical Operating Conditions

	Butene Film Grade	High-Denisty Grade
Melt Index GM/10ALL	1.0	7.0
Density GM/CC	.918	.953
Temperature °F	190	225
Pressure PSIG	310	310
Partial Pressure PSIA		
Ethylene	135	150
Butene	47	7.5
Hydrogen	19	90
Nitrogen	109	62.5
Rate: Klbs/hr	60	65

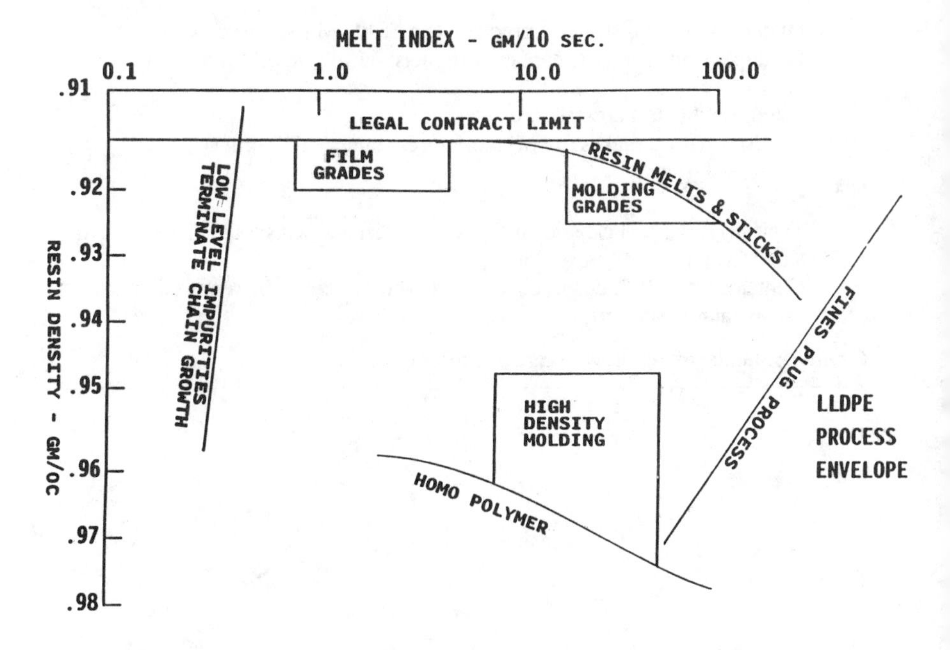

Figure 3-10 Melt index

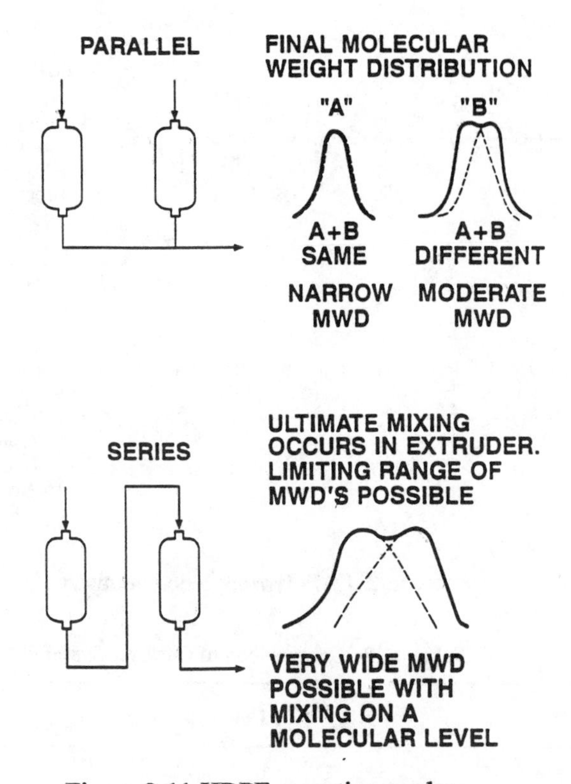

Figure 3-11 HDPE operating modes

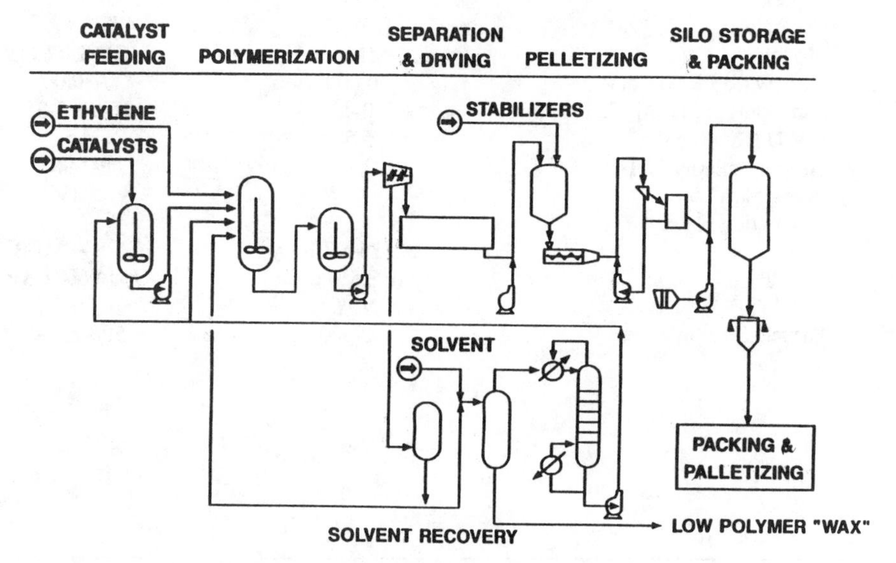

Figure 3-12 HDPE process flow scheme

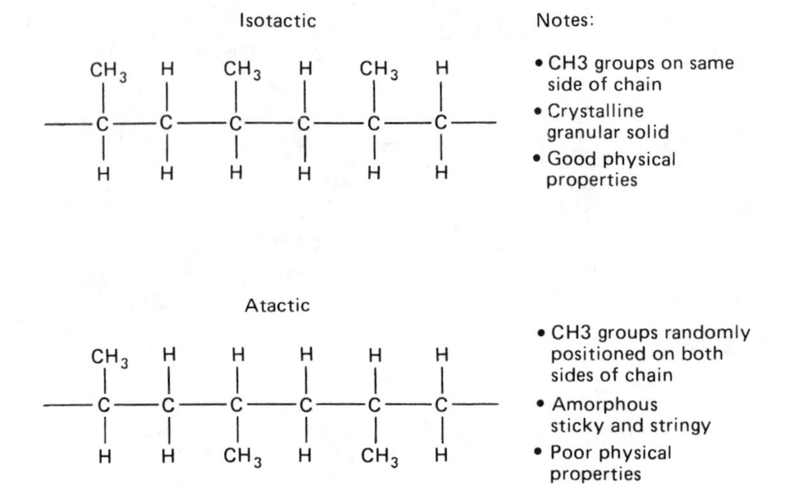

Figure 3-13 Polypropylene chemistry

Table 3-19 Comparison of Catalyst Systems

	Low Aluminum	**Supported**
Catalyst	$TiCl_3$	$TiCl_3$
Support	None	$MgCl_2$
Cocatalyst	DEAC	TEAL
Modifier	MMA	MCMS, TEOS
Activity (lb PP/ lb Cat)	10,000	20-30,000
Ash Content (Ti, Al, Cl, Mg) ppm	10-40	35-70, 5-25
MWD (Mw/Mn)	5.5	4.5
Stereoregularity (%HI)	93-97	92-98
Production Lines	5, 6, 7	5, IV
Production Slate:		
HP	0.1-35 MFR	0.1-800 MFR
RCP	Up to 3.5% C_2=	Up to 6% C_2=
ICP/TPO	N/A	Up to 16% C_2=
Particle Size (microns)	500-800	500-2,000

$TiCl_3$ — $MgCl_2$

Table 3-20 Product Attributes

	Definition	Uses
Homopolymer	Normal unsubstituted polypropylene backbone	Film, Nonwovens, and Fiber
Random Copolymer	Low levels of ethylene randomly dispersed in a copolymer with propylene	Molded food containers and film for the food packaging industry
Impact Copolymer	Homopolymer backbone with a small amount of an EP copolymer (much like rubber) dispersed in it.	Various injection molding application where improved impact strength is needed

Special process functions or requirements of a finishing extruder include:

- High throughput rates >25 tons/hour for LLDPE; 20 tons/hour for (PP/HDPE/LLDPE)
- Minimize melt flow rate (mfr) breakdown or mi (melt index) shift (PP/HDPE/LLDPE)
- Controlled mfr breakdown using oxygen or peroxide (Cr'D PP)
- Minimize gels, off color, black specks (LDPE)
- Homogenization to improve properties (LDPE, ICP, PP)
- Venting (PP)
- Devolatilization (BRPP F LINE LLDPE)
- Blending (ICP PP + LLDPE< ICP PP + HDPE< LLDPE + PIB, PP + LLDPE, LDPE + LLDPE, and so on.)
- Wide grade slate range (0.5 - 800 mfr. bapp, 0.3 -2500 MI app/brpp, 0.5 mi LLDPE to 100 MI HDPE Kemya/MBPP/SARNIA)

Table 3-21 provides a listing of extruder types for powder and granular-type feeds.

Figures 3-24 through 3-33 provide diagrams of various commercial finishing extruders for plastics.

Table 3-22 provides a summary of product forms commercially on the market today, along with general properties.

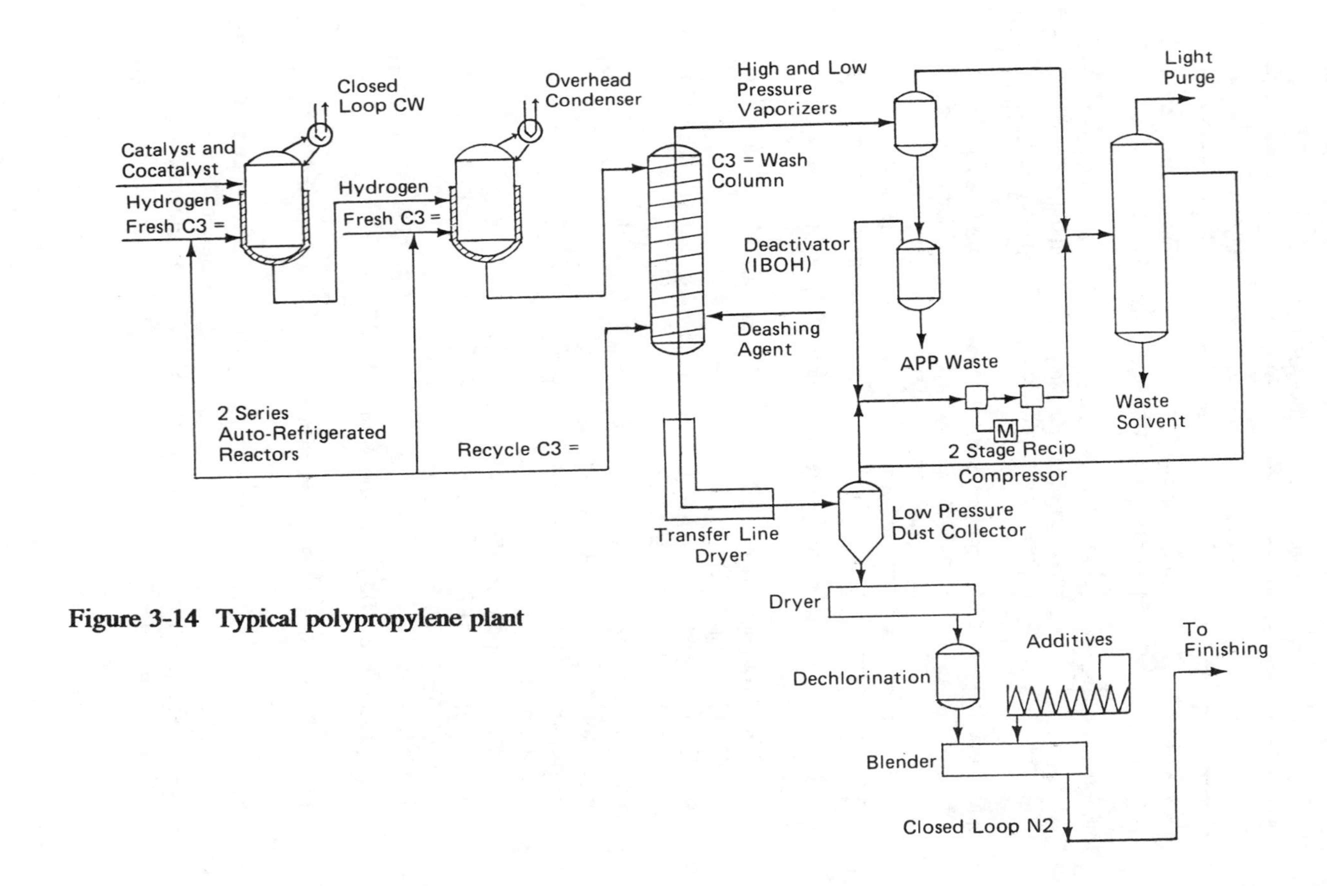

Figure 3-14 Typical polypropylene plant

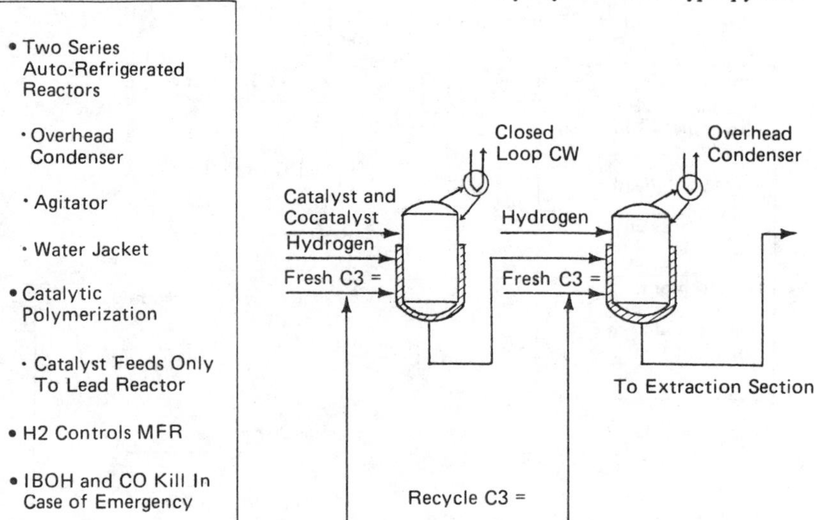

Figure 3-15 Reactor section of polypropylene process

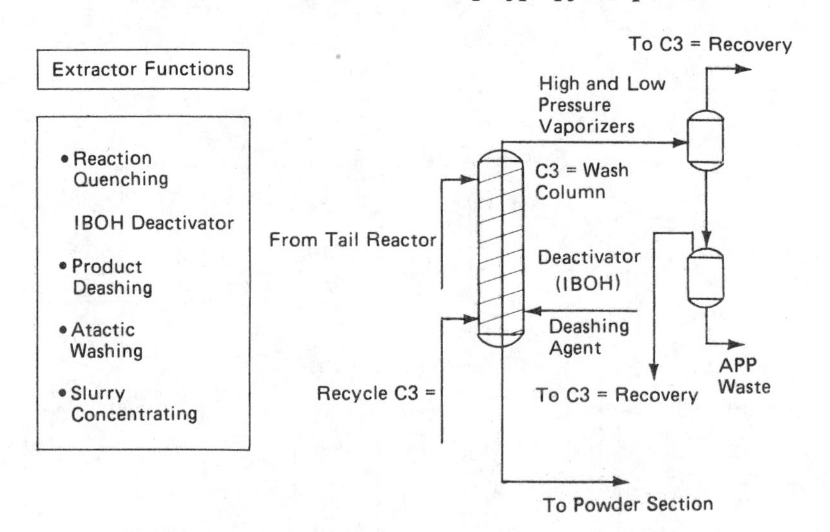

Figure 3-16 Extraction section of polypropylene process

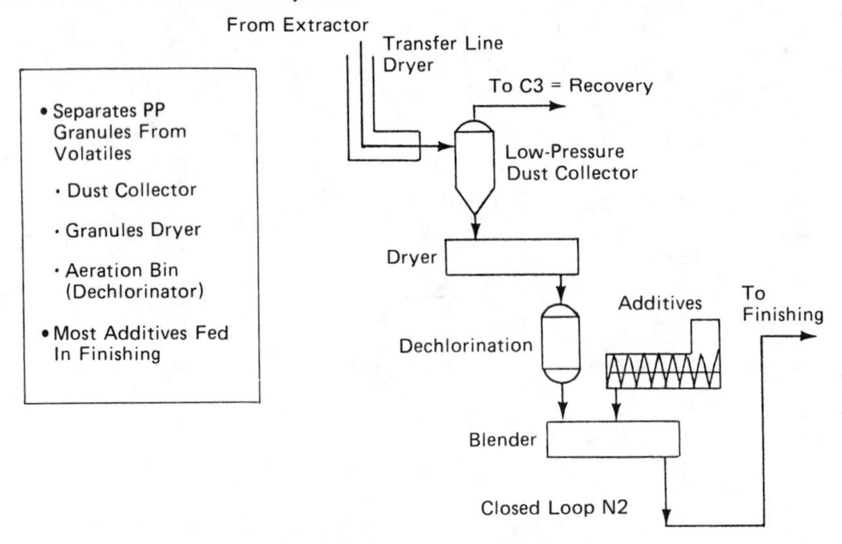

Figure 3-17 Powder handling section of polypropylene process

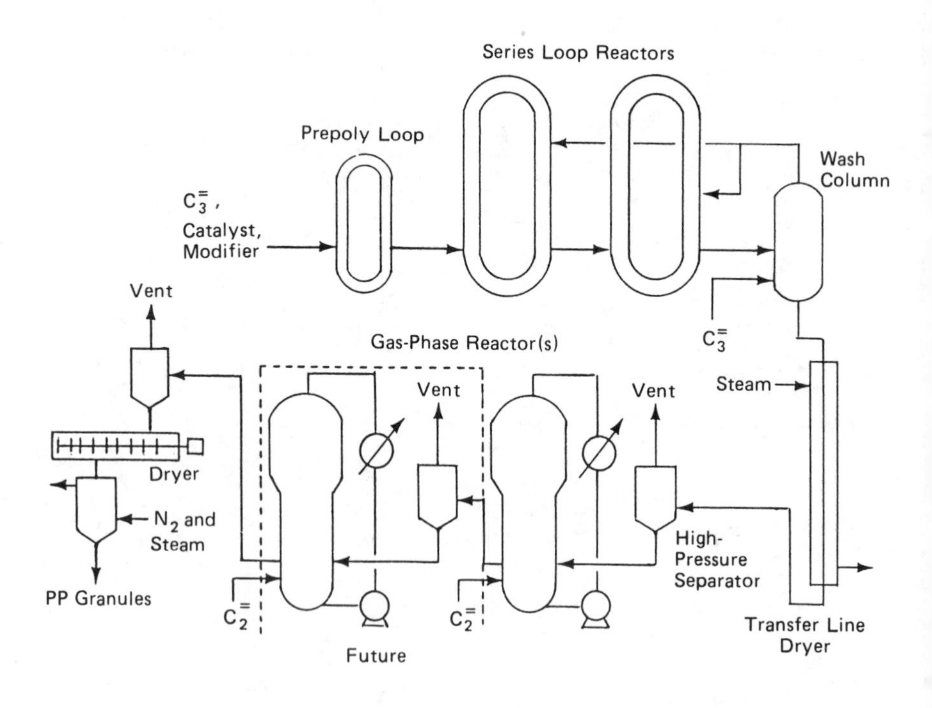

Figure 3-18 Exxon-Himont-Mitsui gas phase process

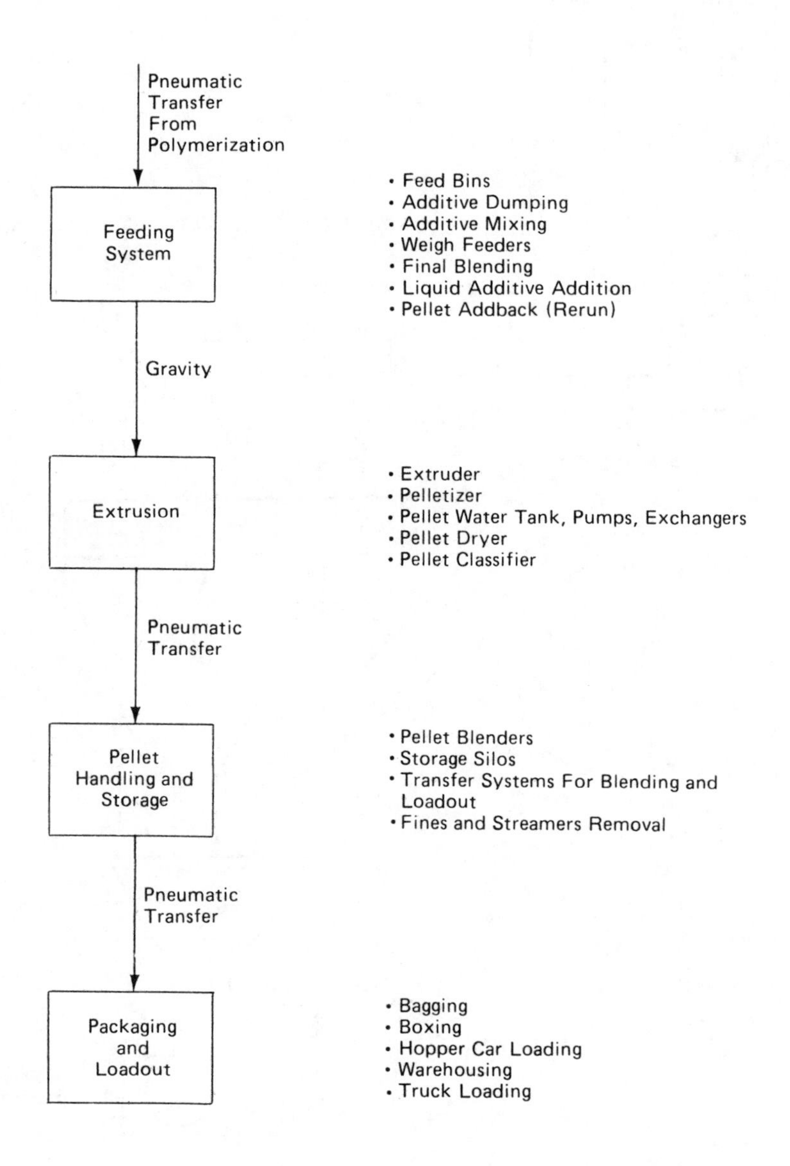

Figure 3-19 Typical finishing steps in plastics finishing

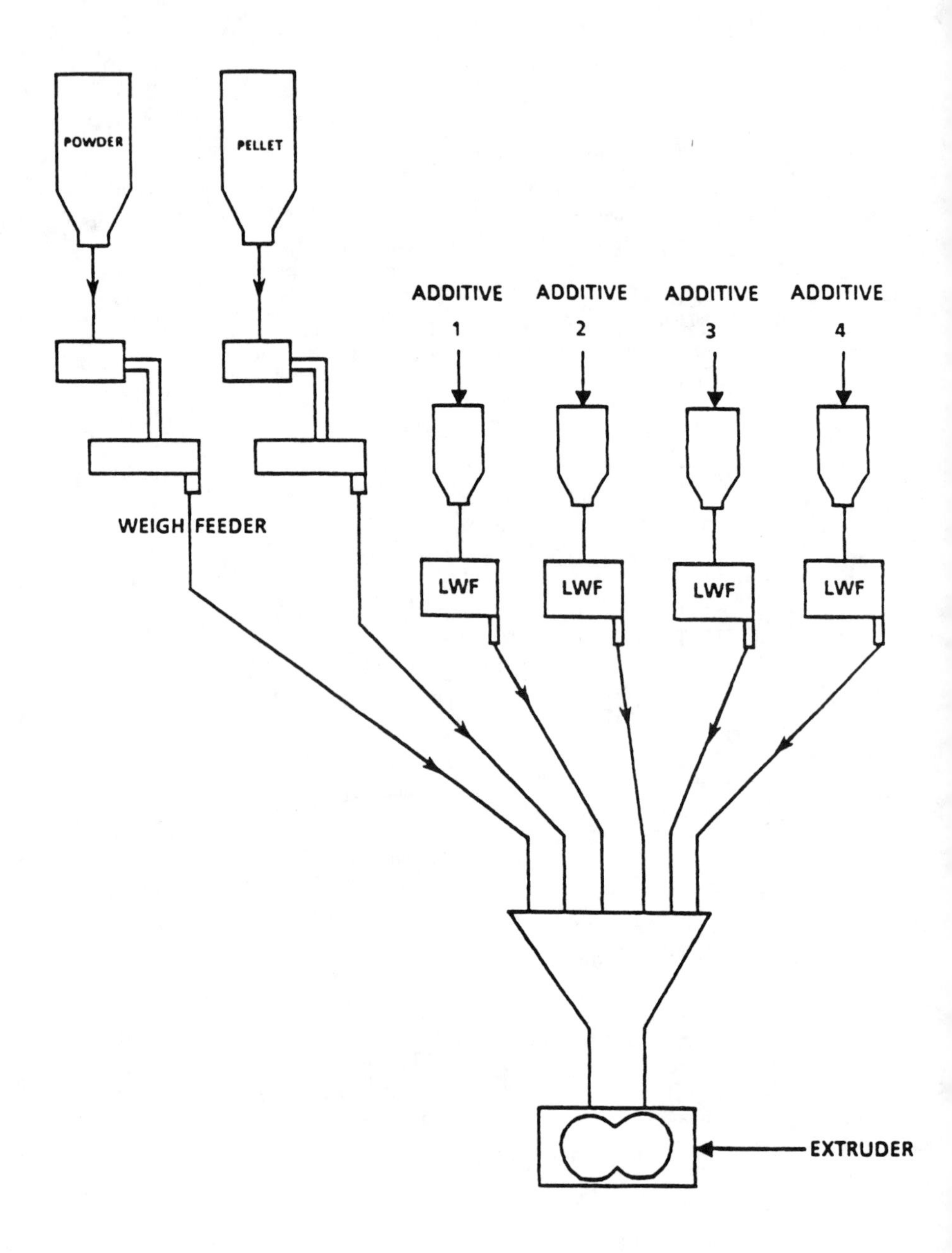

Figure 3-20 Additive addition: Direct feed without preblending

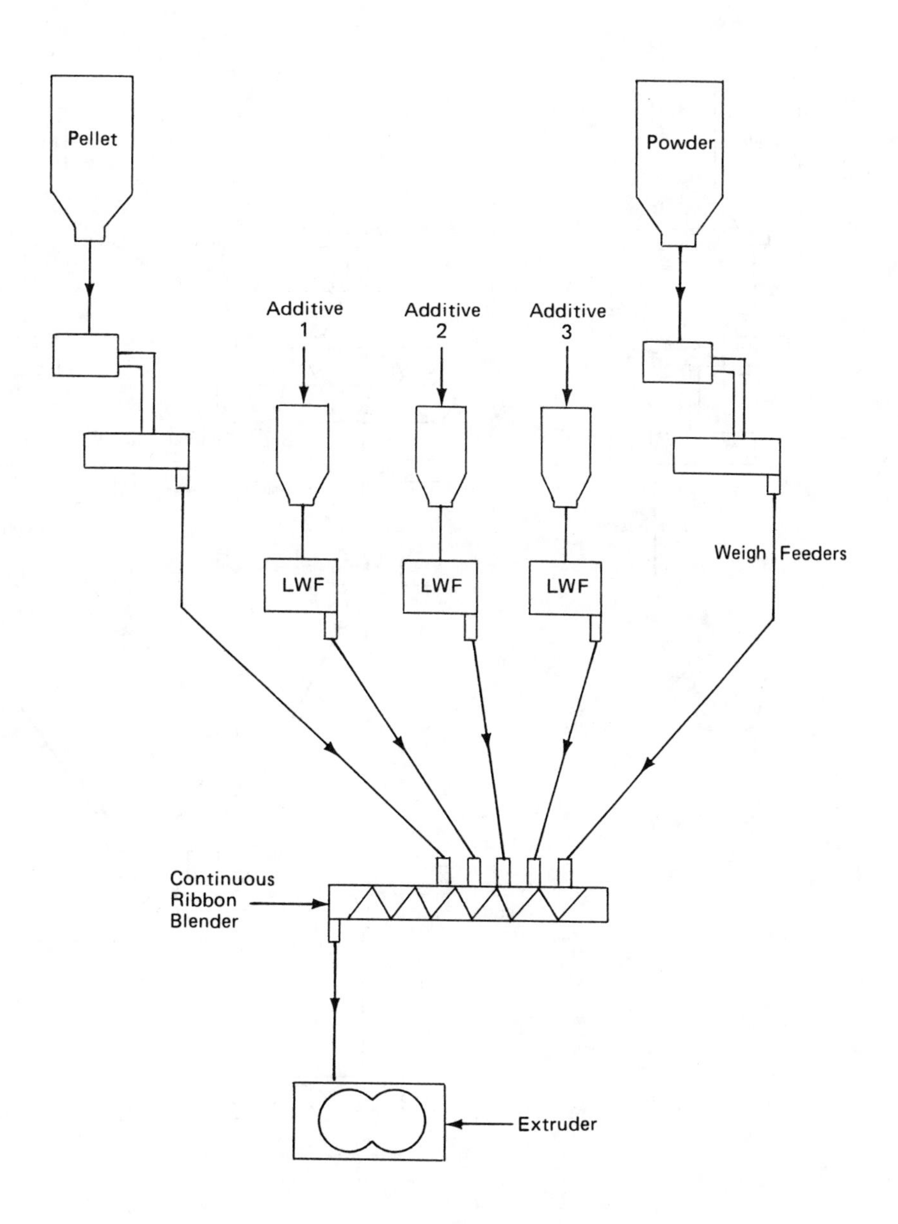

Figure 3-21 Additive addition: Direct feed with preblending

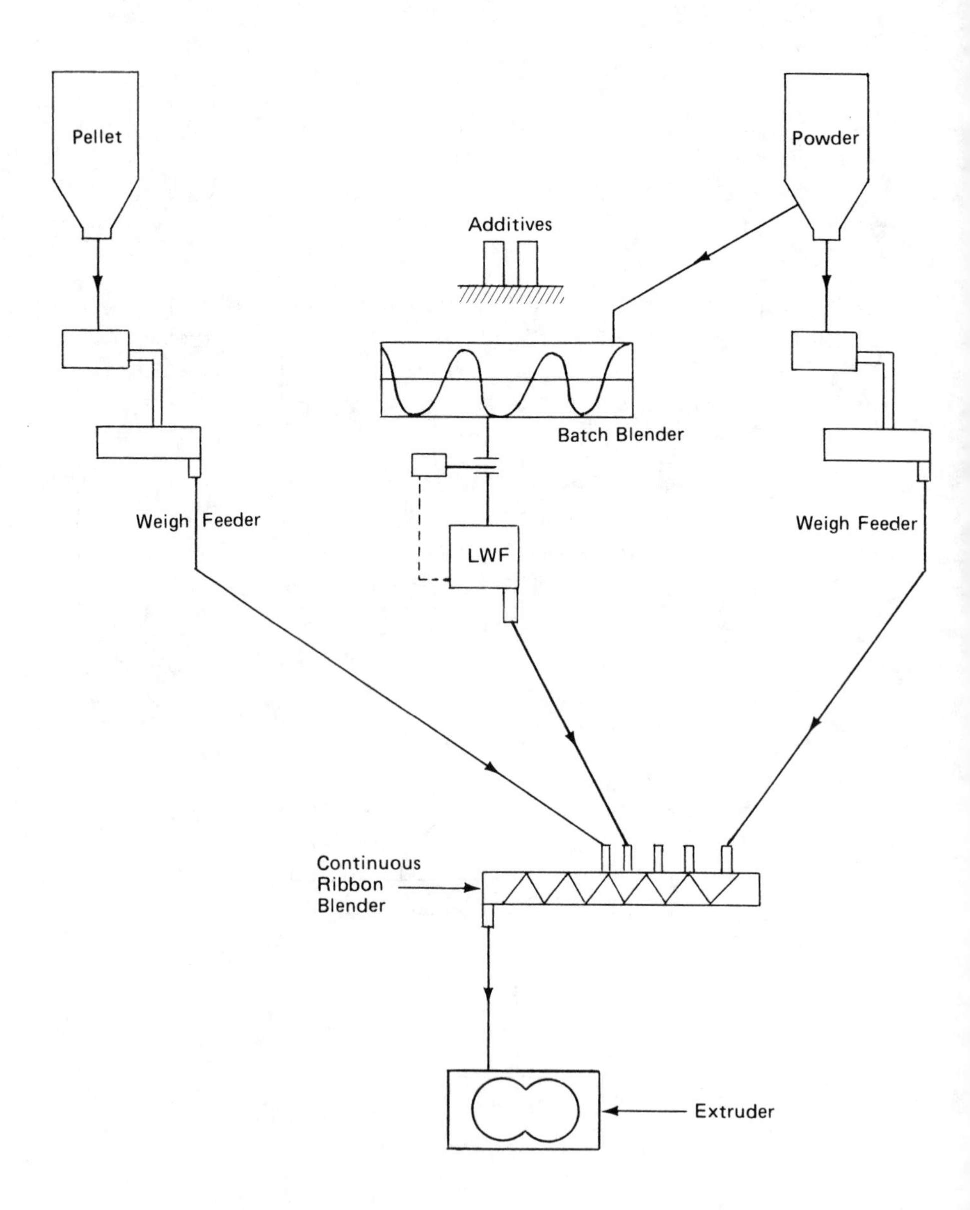

Figure 3-22 Additive addition-Granuks masterbatch

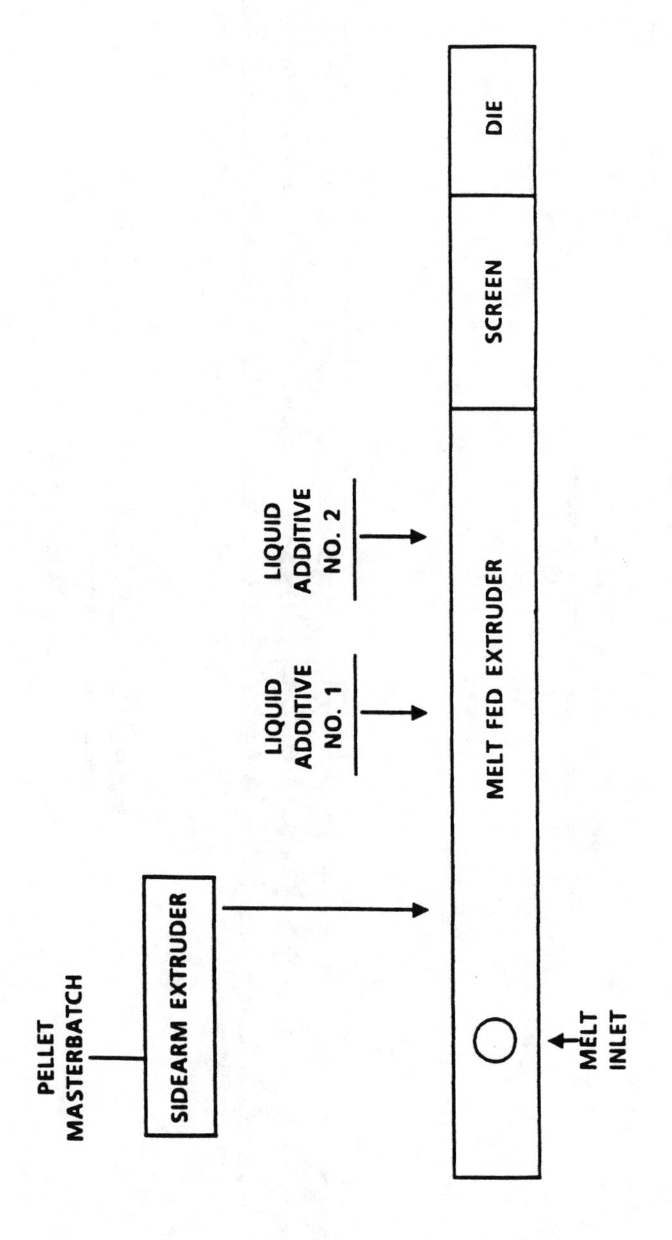

Figure 3-23 **Plastics finishing: additive addition melt-fed extruders**

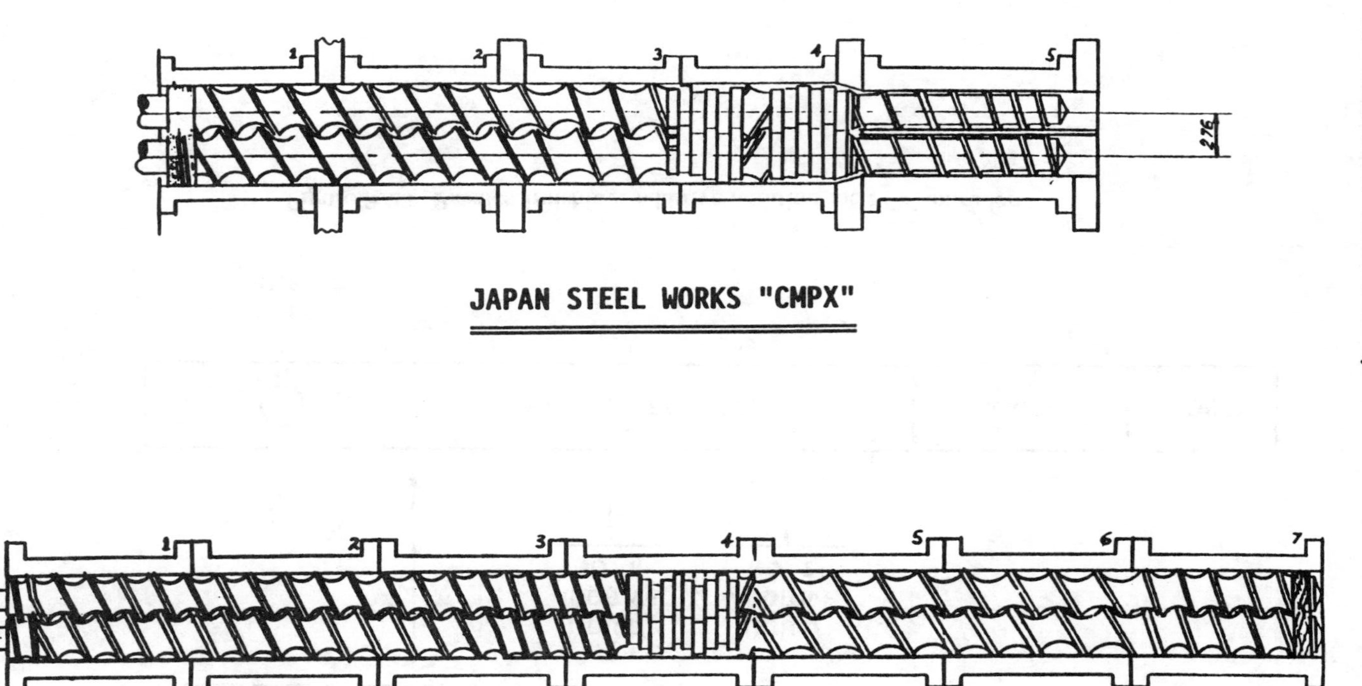

Figure 3-24 Examples of screw designs

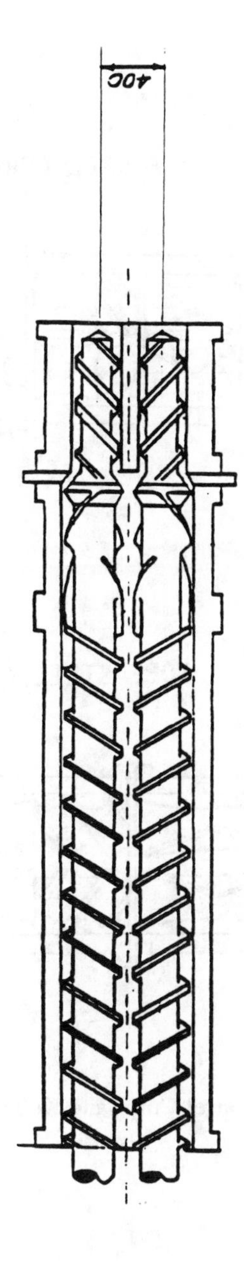

Figure 3-25 Japan steel works counter-meshing twin screw design

FARREL CORP.

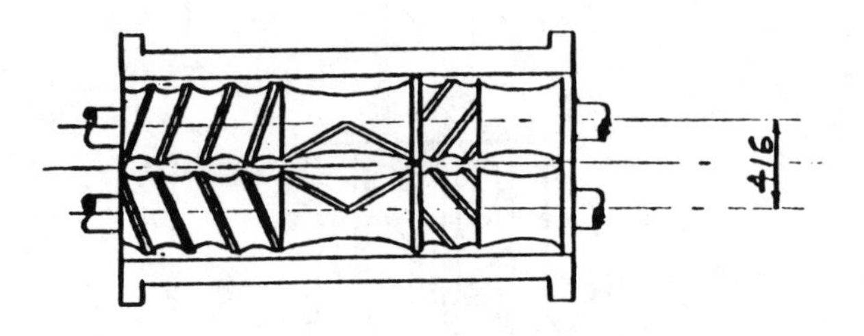

KOBE STEEL CO.

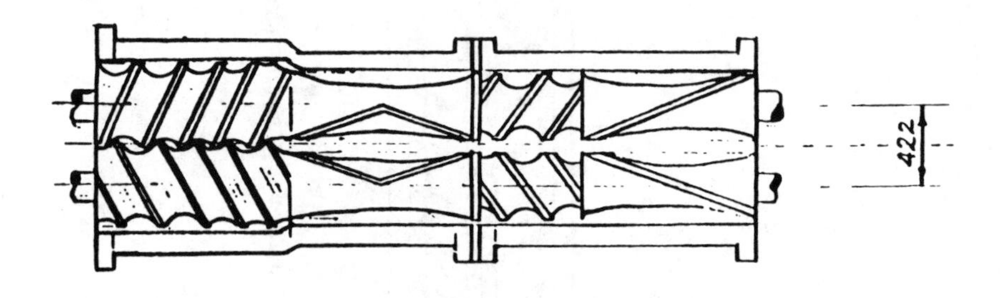

Figure 3-26 Illustrates Farrel Corp. and Kobe Steel Co. screw designs

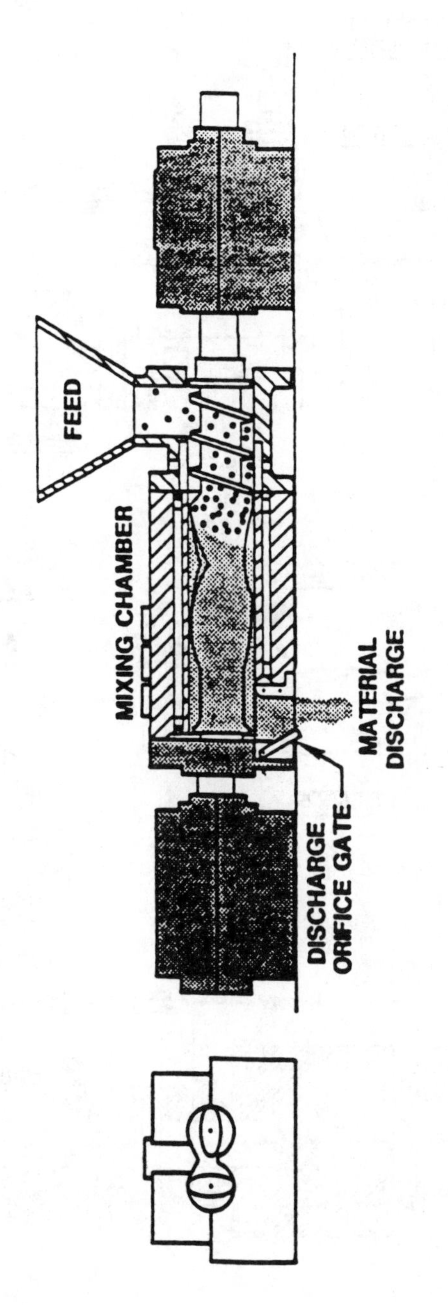

Figure 3-27 Farrel FCM design

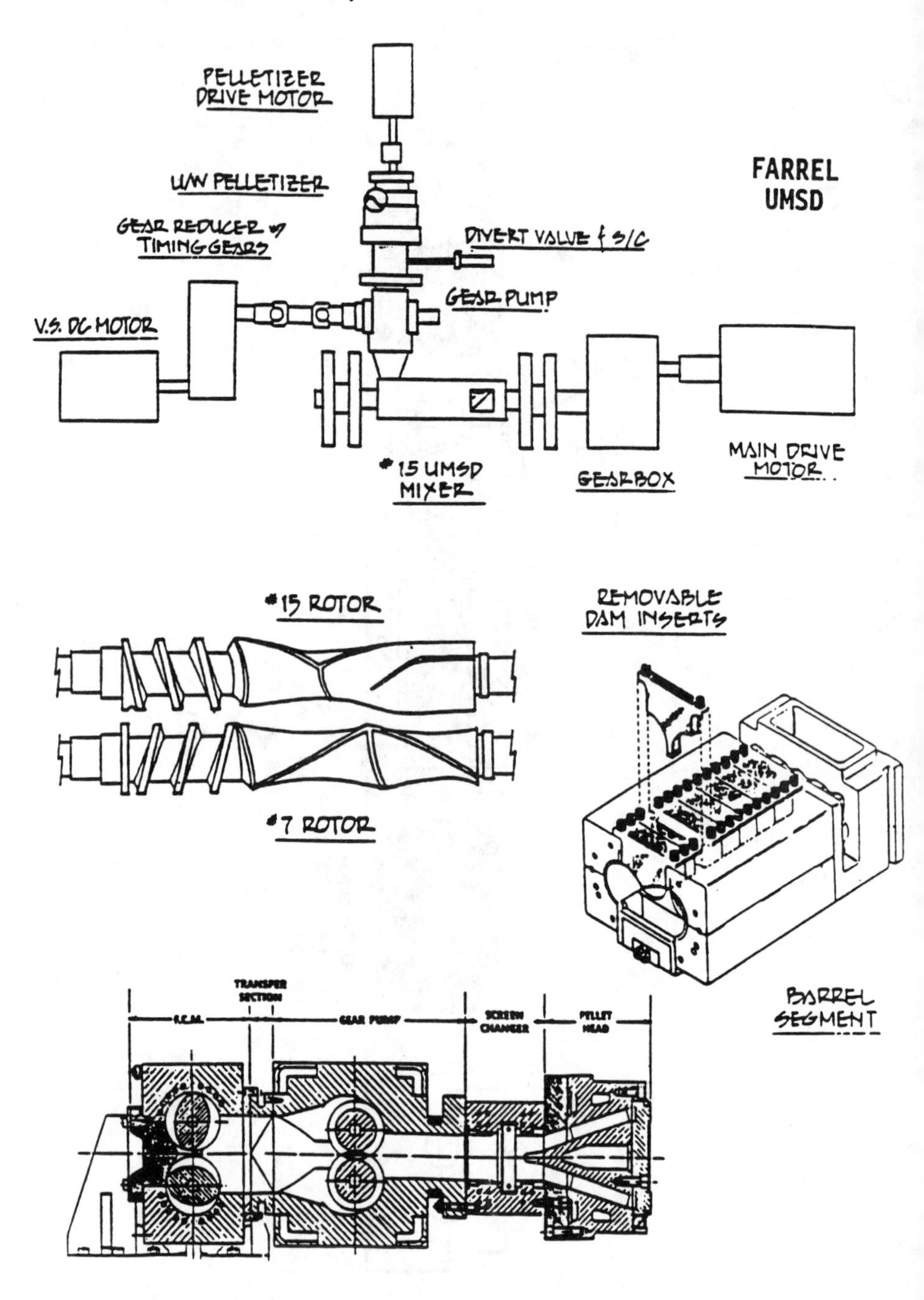

Figure 3-28 Examples of Farrel designs

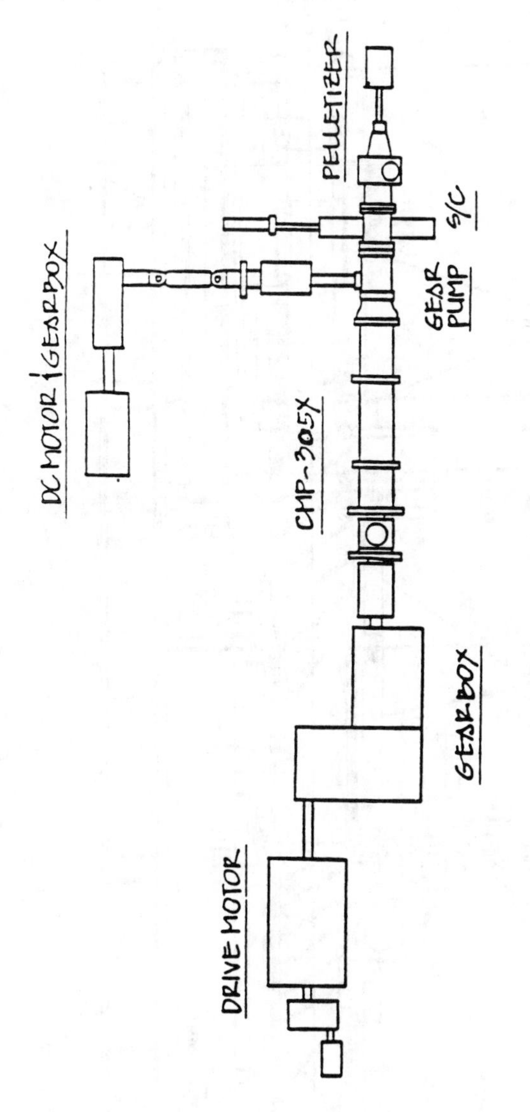

Figure 3-29 Japan steel works CMPX

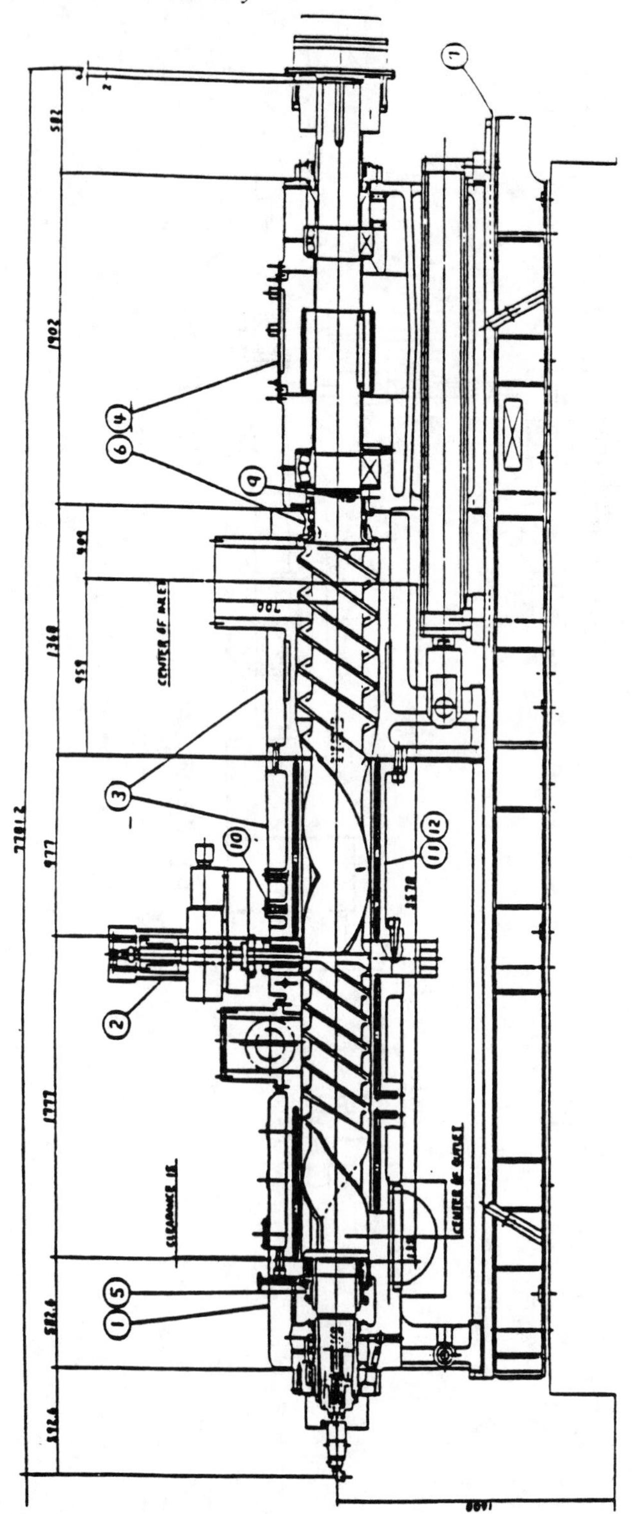

Figure 3-30 Kobelco: LCM-380G system

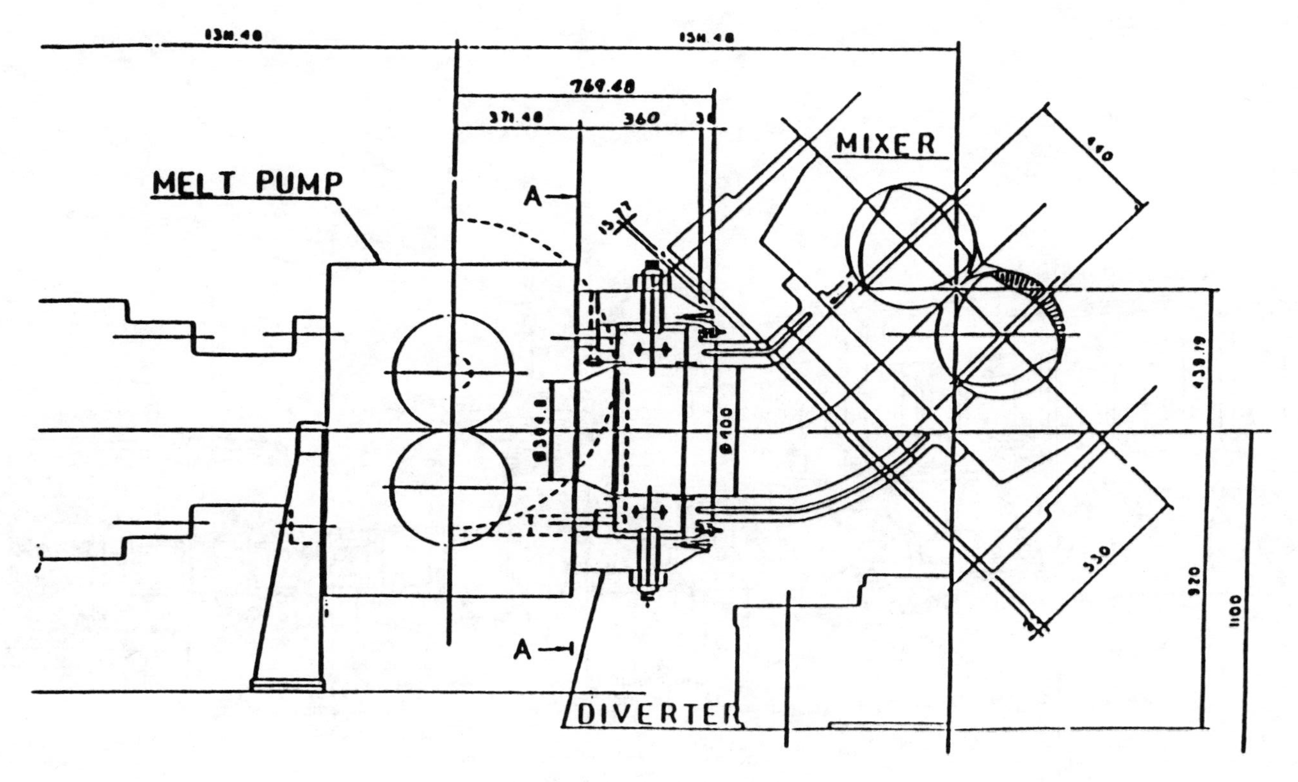

Figure 3-31 Kobelco design

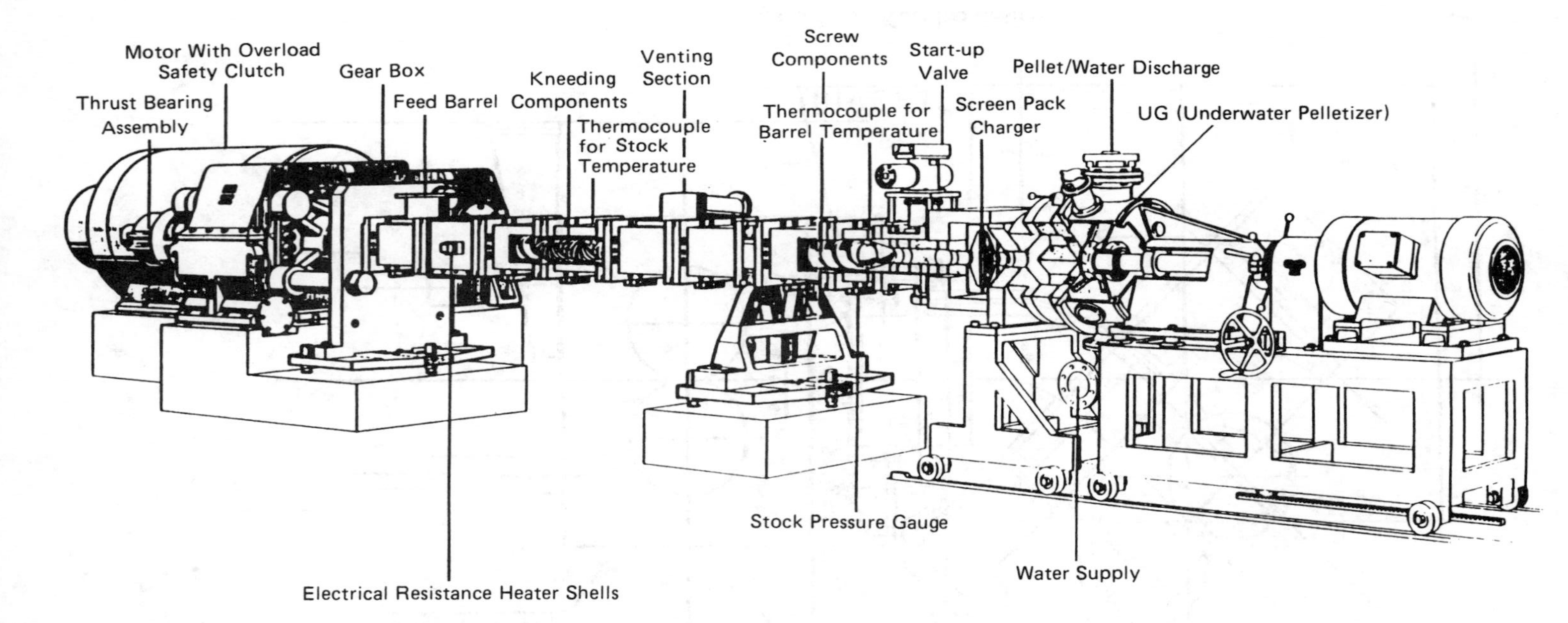

Figure 3-32 Werner & Pleiderer system

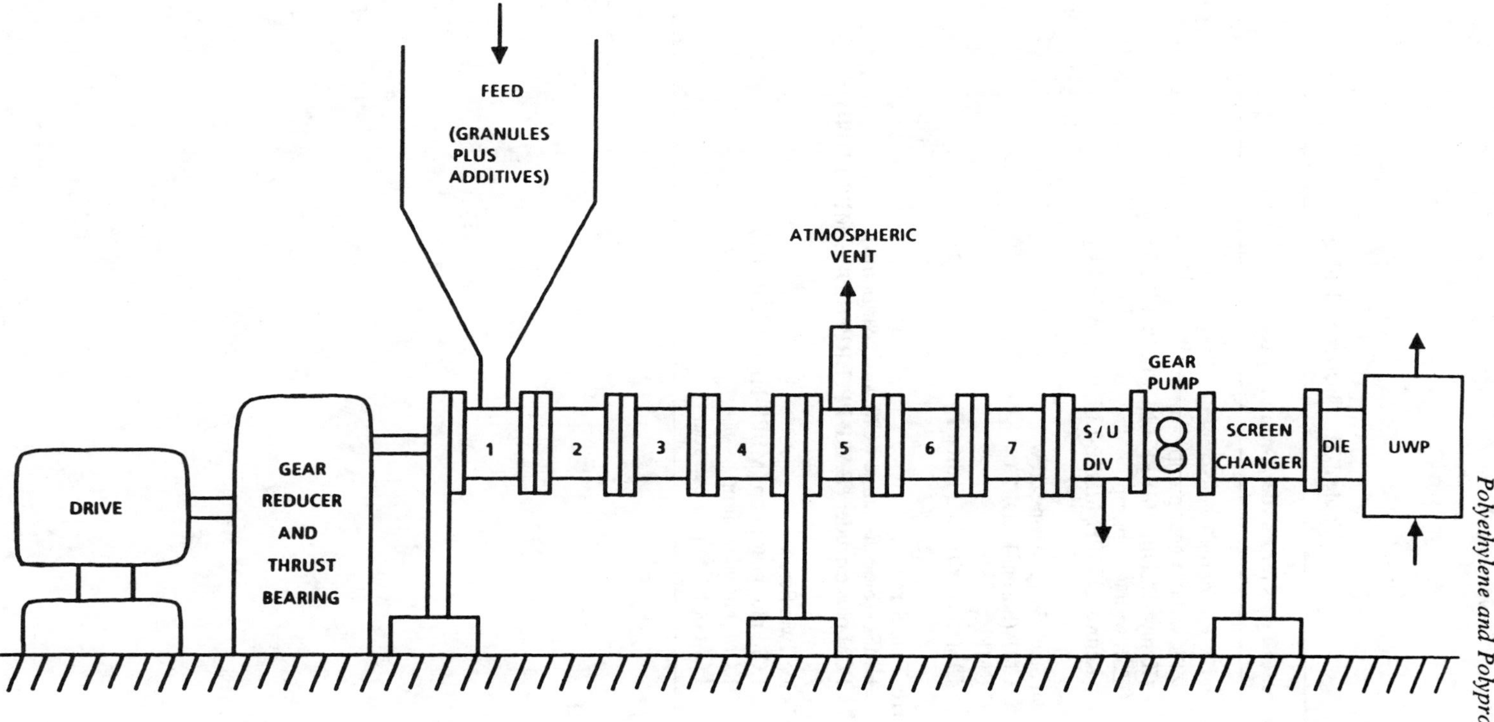

Figure 3-33 Typical finishing extruder

Table 3-21 Plastics Finishing

Extruder Types for Powder/Granular Feeds
Werner & Pfleiderer Twin Screw
ZSK without gear pump
(Kemya-LLDPE, MBPP-LLDPE, BAPP-PP)
ZSK with gear pump
(Kemya-LLDPE, MBPP-LLDPE, BAPP-PP)
Japan Steel Works Twin Screw
CMP (Sarnia-LLDPE)
CMPX
CIM plus extruder (TSK-PP)
Farrel Twin Screw
UMSD (Side discharge FCM with gear pump)
FCM plus extruder (BAPP-PP, BRPP-LDPE, MPP-LDPE)
Kobelco Twin Screw
LCM plus gear pimp (MBPP-HDPE, TSK-PP)
KCM plus gear pump
LCM or KCM plus extruder

Table 3-22 Plastics Finnishing - Product Form Can Influence Process Design

Type	Description	Average Particle-Microns	Particle Distribution	Bulk Denisty lbs/ft^3	Example Products	Example Location	Other notes
Granule	Free Flowing	800	Narrow	32-34	HP, RCP	BAPP, TSK	
Granule	Free Flowing	600	Narrow	28-32	HP, RCP, ICP	BAPP, TSK Line 3	Mitsui T-Catalyst
Granule	Free Flowing	1,000-3,000	Broad	28-30	HP, RCP, ICP	BAPP Line IV	Himont Spheripol Catalyst
Granule	Free Flowing	650	Broad	27-29	LLPDE	MBPP, Kemya, Sarnia	Butene Copolymers
		770	Broad	22-25	LLPDE	MBPP, Keyma, Sarnia	Hexene Copolymers
Powder	Fluidizable	80-100	Broad	17-21	HDPE	MBPP	MPC Process (1990) S/U)
Powder	Fluidizable	100	Broad	25	PP	TSK-Line 1	
Melt	–	–	–	45	LDPE	APP, MPP, BRPP	
	–	–	–	45	LLPDE, VLDPE	BRPP F-Line	Up to 3% Volatiles

4 *ETHYLENE-PROPYLENE ELASTOMERS*

4.1 OVERVIEW

Ethylene-propylene synthetic rubbers are classified by the ASTM and ISO as synthetic elastomers with an "M" designation, meaning that they have a saturated polymer chain of the polymethylene type. Within this classification there are two basic kinds of rubber: EPM, the copolymer of ethylene and propylene; and EPDM, the terpolymer of ethylene, propylene, and a nonconjugated diene with residual unsaturation in the side chain. The diene is introduced to allow conventional sulfur vulcanization; the location on a side chain prevents the reactive unsaturation from being a site for molecular breakdown.

This class of elastomer was first introduced in Europe in the early 1950s utilizing the catalyst technology for polymerizing α-olefins discovered by Ziegler and advanced by Natta. Both researchers were awarded the Nobel Prize for Chemistry in 1963. The original process involved the polymerization of two simple olefins, ethylene and propylene, in the presence of the Ziegler catalyst, at approximately ambient temperature and atmospheric pressure, resulting in an amorphous-to-semicrystalline elastic material. Manufacturing processes fit into two categories: the solution process and the slurry process.

After commercial introduction, interest in this hydrocarbon elastomer intensified because of two unique features:

1. A completely saturated polymer backbone, which provides virtually unlimited ozone and weather resistance, good heat resistance, low-compression set, and low-temperature flexibility
2. An ability to accept relatively large quantities of low-cost filler and oil compared to other rubbers, while retaining a high level of compounded physical properties

These properties have led to a large number of consumer-oriented applications. An overview of the end-use properties and processing operations used for these materials as well as other types of rubbers is presented in this chapter.

4.2 GENERAL PROPERTIES

Ethylene-propylene polymers are inherently stable rubbery materials, off white to amber in color, and somewhat translucent. Although stable in normal storage, these materials, especially oil-extended polymers and very high-diene polymers, must be protected from light-induced cross-linking before compounding and use. Typical physical constants of raw EPM/EPDM polymers before compounding are given in Table 4-1.

Table 4-1 Typical Physical Characteristics of Raw EPM/EPDM Polymers

Property	Value
Specific gravity	0.86 to 0.99
Heat capacity, J/(g°K)	2.22
Thermal conductivity, kW/(m-K)	1757
Thermal diffusivity, cm/s	1.9×10^{-3}
Thermal coefficient of linear expansion, 1/°C	2.5×10^{-4}
Brittle point, °C	-55 to -65
Glass transition temperature, °C	-55
Air permeability, cm^2/s-atm	100

Although compounding with carbon black and other ingredients plays a dominant role in processing and end-use requirements, the characteristics of the base polymer also establish final performances. The principal polymer characteristics of importance are:

Mooney viscosity (molecular weight)
Ethylene content (low-level crystallinity)
Diene content
Molecular weight distribution
Physical form

In terms of expected performance, compound properties best describe a rubber product for the applications engineer. Each compounding ingredient must be carefully selected to enhance the processing and performance properties of the base polymer in the compound. Properly formulated, the ethylene-propylene polymer molecule imparts these significant features to products made from it:

Outstanding resistance to ozone attack
Excellent weathering ability

Excellent heat resistance
Wide range of tensile strength and hardness
Excellent electrical properties
Flexibility at low temperatures
Good chemical resistance, especially to polar media
Resistance to moisture and steam

Typical ethylene-propylene rubber compound properties are given in Table 4-2. Much of the environmental stability results from the fact that the side-chain double bonds, although reactive, do not act as sites for polymer chain breakdown. Table 4-3 compares EPDM properties to those of other commonly used rubbers containing main-chain double bonds. E-P rubbers are practically impervious to ozone degradation. Vulcanizates are unaffected under the customary test conditions of 100 pphm ozone and 20 percent extension. Even after 15 months exposure to this concentration and strained up to 60 percent, no cracking can be observed. When used in black compounds, this inherent resistance to ozone combined with superior resistance to degradation by ultraviolet radiation and extremes of atmospheric temperatures makes this rubber ideally suited for outdoor use. These properties are obtained without the use of antiozonants or waxes which natural rubber, styrene-butadiene rubber, and chloroprene rubber compounds often require.

When properly compounded, EPM/EPDM vulcanizates are suitable for continuous service up to 150°C with extrusions up to 175°C. This performance is attributed to the completely saturated polymer backbone and the isolation of the pendant olefinic sites from the main polymeric chain.

At high service temperatures, changes in tensile strength of EPM/EPDM vulcanizates are minimal. The restriction on service life is usually caused by loss of elongation after aging at high temperatures. In EPDM, free-radical vulcanization with its carbon-carbon cross-links gives better heat aging than the best monosulfidic sulfur cross-links. Further, as shown in Figure 4-1, the rate of elongation loss in EPM is better than in EPDM because there is no pendant unsaturation, which is considerably more reactive to free-radical oxidative attack than tertiary carbons on the main chain. In addition, the more rubber-rich compounds give the best results, since there is less localized strain amplification acting on the network.

Another important property is chemical resistance. These rubbers have excellent resistance to acids, alkalis, and hot detergent solutions. EPM/EPDM is also resistant to salt solutions, oxygenated solvents, synthetic hydraulic fluids, and animal fats. Table 4-4 gives the results of immersing a typical EPDM compound into different chemical solutions. Chemical resistance was determined according to ASTM D471. The compound tested contained 25 percent RHC (rubber hydrocarbon).

Table 4-2 Typical Properties of EPM/EPDM Compounds

ASTM D-2000 classification	BA, CA, DA, BC, BE
Mechanical properties (reinforced)	
Hardness, Shore A	35 to 90
Tensile strength,MPa	4 to 22
Elongation, %	150 to 800
Tear Strength, kN/m	15 to 50
Compression set	15 to 35
Electrical properties	
Dielectric constant	2.8
Power factor, %	0.25
Dielectric strength, kV/mm	26
Volume resistivity, Ω • cm	1×10^{16}
Thermal properties	
Brittleness point, °C	-55 to -65
Minimum for continuous service, °C	-50
Maximum for continuous service, °C	150
Maximum for intermittent service, °C	175
Maximum theoretical temperature to break hydrocarbon bonds, °C	204
Heat capacity, J/kg • K	
40 GPF/40 oil	1950
160 GPF/100 oil	1790
Dynamic Properties	
Resilience, % (Yerzley)	75
Elastic spring rate (15 Hz), kN/m	550
Loss tangent (15 Hz), %	0.14
Chemical resistance	
Weather	Excellent
Ozone	Excellent
Radiation	Excellent
Water	Excellent
Acids and alkalis	Excellent to good
Aliphatic hydrocarbons	Fair to poor
Aromatic hydrocarbons	Fair to poor
ASTM oils	Good to fair
Oxygenated solvents	Good
Animal and vegetable oils	Fair
Brake fluid (nonpetroleum)	Excellent
Glycol-water	Excellent

Table 4-3 Comparison of EPDM to Other Rubbers and Typical Compounds[a]

Rubber type	EPDM	Natural	SBR	IIR	CR
Specific gravity (polymer)	0.86	0.92	0.94	0.92	1.23
Tensile strength (max.), MPa	22	28	24	21	28
Elongation, %	500	700	500	700	500
Top operating temperature, °C	150 to 175	75 to 120	75 to 120	120 to 180	90 to 150
Brittleness point, °C	-55 to -65	-55	-60	-45	
Compression set, % 22 h at 100°C	10 to 30	10 to 15	15 to 30	15 to 30	15 to 30
Resilience, % (Yerzley)	75	80	65	30	75
Tear Strength, kN/m	15 to 50	35 to 45	25 to 35	25 to 35	35 to 45
Dielectric strength, kV/mm	20 to 55	16 to 24	24 to 32	24 to 36	16 to 24
Resistance to:					
Weathering	E	F-G	F-G	E	G
Ozone	E	P	F	G	G
Acids and alkalis	G-E	G	G	G	E
Oils and solvents	P-F	P	P-F	P-F	G
Abrasion	G-E	G-E	G	G	G-E
Compression set	G-E	E	G	G	G
Tearing	G	E	F	G	G-E
Low temperature	G-E	G	F-G	G	F
Steam	G-E	P	P	E	G
Air permeation	F	F	F	E	F

[a] SBR, Styrene-butadiene rubber; IIR, isobutylene-isoprene rubber (butyl rubber); CR; chloroprene rubber (neoprene); E, excellent; G, good; F, fair; P, poor.

Although ethylene-propylene rubber compounds have only limited resistance when immersed in hydrocarbon solvents such as toluene and gasoline, they can be useful in a hydrocarbon atmosphere when exposure is intermittent or mild. This is evidenced by its successful use in automotive underhood applications.

Other properties of importance include low-temperature performance and dynamic behavior. The resilience at room temperature is slightly less than that of natural rubber and generally equivalent to styrene-butadiene and polychloroprene elastomers. Although the low-temperature brittle point of ethylene-propylene rubber is about the same as that of styrene-butadiene rubber, it retains a greater percentage of its resilience at low temperatures. Figure 4-2 shows this in terms of stiffness. The dynamic response of E-P rubber compounds is somewhat similar to that of natural rubber compounds, as indicated in the Yerzley test comparison illustrated in Figure 4-3. Ethylene-propylene rubber, however, is a more popular choice for dynamic parts because its age resistance better preserves initial design characteristics with time and environmental extremes. EPDM is a good first choice when high resiliency is desired. Where high damping is called for, butyl rubber is the proper choice, as shown in the Yerzley test.

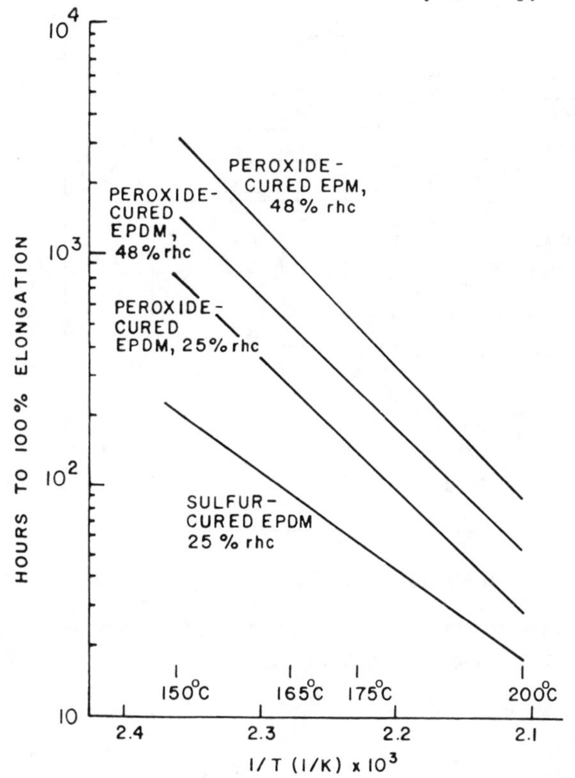

Figure 4-1 Typical data on heat resistance of EPM/EPDM

Table 4-4 EPDM Chemical Resistance, 72-Hour Immersion at 100°C

	Volume change (%)	Tensile strength retained (%)	Elongation retained (%)	Hardness change (points)
ASTM No. 1 oil	+62	103	94	-30
ASTM No. 3 oil	+108	60	71	-42
Dioctyl phthlate	+7	107	89	-3
Distilled water	+1	112	84	0
Hexane[a]	+55	55	53	-25
Hydrochloric acid (10%)	+14	116	91	-3
Methyl ethyl ketone[a]	-5	72	68	+5
Perchloroethylene	+73	66	66	-35
Skydrol 500 brake fluid	-5	110	65	+5
Sodium chloride (25%)	0	118	84	-1
Sodium hydroxide (10%)	+2	121	92	+1
Detergent (1%)	+1	115	91	+1
Toluene	+92	47	53	-40

[a] 72-hr immersion at boiling point of test liquid.

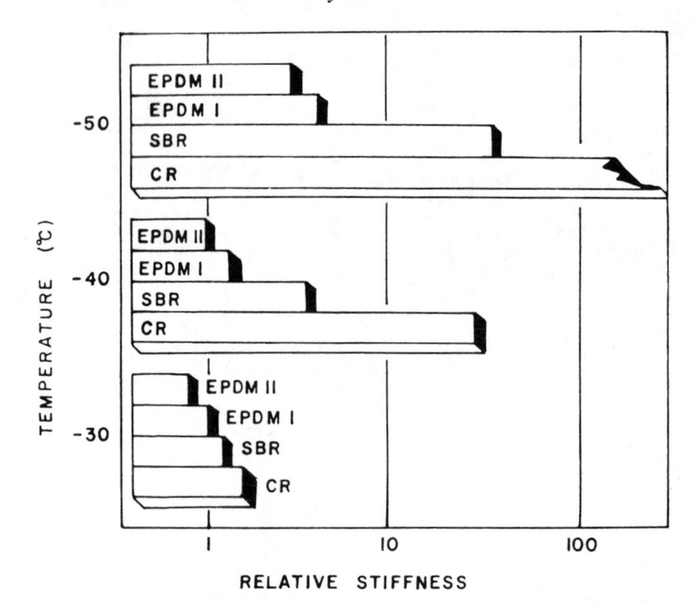

Figure 4-2 Comparison of relative stiffness in torsion of different rubbers EPDM I, crystalline; EPDM II, amorphous

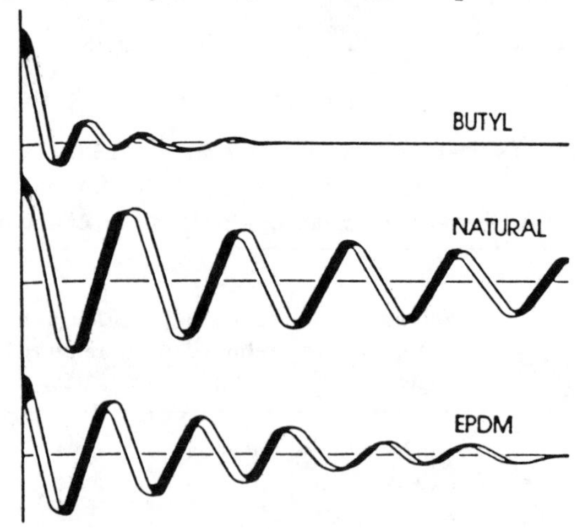

Figure 4-3 Yerzley test for different elastomers

4.3 COMPOUNDING AND VULCANIZATION

Rubber compounds contain a number of ingredients; generally, these fall into five categories: (1) elastomers, (2) fillers, (3) plasticizers, (4) miscellaneous chemicals, and (5) vulcanization system. Selection and application of these to the compound are determined by end-use requirements and ultimately represent a balance of the following four parameters.

1. Vulcanization (cure) rate
2. Vulcanizate physical properties
3. Cost
4. Processing behavior

As a general rule, ethylene-propylene elastomers are amorphous polymers. Unless the molecules are above approximately 60 wt percent ethylene, they do not develop crystallization on stretching and therefore must be reinforced to achieve useful properties. Significantly large volumes of carbon black and other reinforcing fillers and oil can be incorporated—100 to 400 parts per hundred of rubber are common. This in addition to the low specific gravity of ethylene-propylene rubber compared to other rubbers allows compound cost to be kept low.

In many cases, property requirements of the application and preference for sulfur cures will lead the polymer selection toward a terpolymer instead of a copolymer. However, certain performance requirements, such as extreme heat resistance or optimum cost in uncured applications, may direct the choice toward copolymers. Especially with terpolymers, there are also situations where a blend of polymers will be best for a particular application.

The polymer or polymer blend finally selected will influence, to a great extent, the processing characteristics of the compound, the level of physical properties achieved, and to a lesser extent, the compound cost. Property and processing benefits and limitations from increasing molecular weight, ethylene content, diene content, and molecular weight distribution and breadth are summarized in Table 4-5.

A brief description of the major neat polymer properties follows. To begin, neat polymer Mooney viscosity is a bulk viscosity measured in a shearing-disk viscometer. The instrument is illustrated in Figure 4-4. There is no simple mathematical relationship between Mooney viscosity and any of the standard molecular weight averages for polymers. This is partly because of the deliberate changes in molecular weight distribution from polymer to polymer. However, higher Mooney viscosity is a general indication of higher molecular weight. Commercial EPM and EPDM products fit into categories from low to very high Mooney viscosity. In general, the following compound performance observations in terms of processing and vulcanizate properties are made with increasing molecular weight.

Processing effects:

Increased hot green strength (improved collapse resistance; improved low pressure, continuous cures)
Potential for poor dispersion in mixing
Slower extrusion and poor calenderability vulcanizate properties
Increased tensile and tear
Lower compound cost as filler-loading capability increases

Table 4-5 Benefit Summary of EPDM Selection

	As property increases:	
Property	**Benefit**	**Limitation**
Molecular weight (Mooney viscosity)	Tensile and tear strength increases	More difficult to disperse
	Black/oil loading can increase for lower cost; hot green strength increases	Extrusion rate and ability to calender decrease
	Collapse resistance improves	
	Lower-pressure CVa cures improve	
Ethylene content	Cold green strength increases greatly	Tougher to mix
	Flow at high temperature increases	Poorer low-temperature set and flex
	Tensile strength increases	Higher tension set, less elastic recovery
	Higher filler/oil loadings possible	Higher Shore hardness
	Easier to pelletize raw polymer and compounds	
	Peroxide cures improve	
Diene content	Faster cure rate	Compound cost may increase
	Better low-pressure CV cures	Scorch safety decreases
	More flexibility in selecting accelerators	Shelf life decrease
	Compression set improves	Elongation decreases
	Modulus increases	
Molecular weight distribution		
Narrow MWD	Faster cure	Lower green strength
	Faster extrusion speed	Poorer extruder feed
	Better extrusion smoothness	Poorer mill handling, adhesion to mill roll
	Low die swell	Chance of extrusion melt
fracture		
		Fracture calendering
Broad MWD	Improved calendering	Higher die swell
	Improved mill handling	Slower cure rate and lower cure rate
	Higher green strength, especially when hot	Chance of rougher extrusion face
	Improved extruder feeding at all temperatures	
	Improved collapse resistance	
	Less compound flow and tack	

[a] CV, continuous vulcanization.

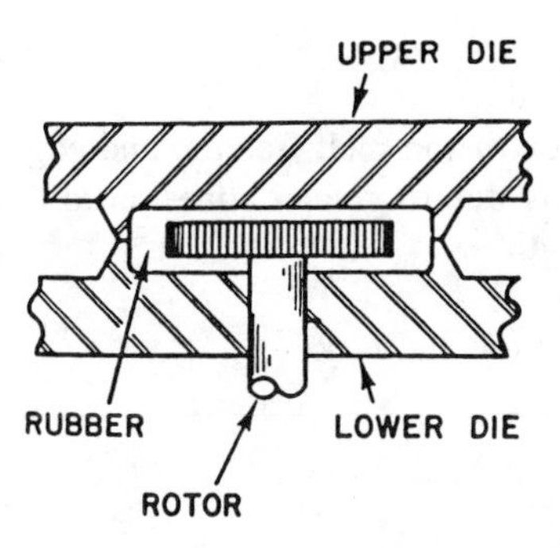

Figure 4-4 Mooney viscometer chamber and rotor

Increasing ethylene content gradually introduces low-level crystallinity above 55 to 65 percent ethylene. This crystallinity usually melts in the range of 30° to 90°C. Therefore, it has a considerable effect on the processability at temperatures below about 90°C; the upper temperature limit varies with molecular weight distribution also. General processing and vulcanizate effects with increasing ethylene content are as follows:

Processing effects:

Increased cold and warm green strength
Easier pelletization of both polymer and compound (important in some feeding operations in end users)
Increased difficulty in mixing and dispersion

Vulcanizate properties:

Increased tensile and crystallinity
Higher hardness
Lower compound costs through higher filler-loading capability
Improved peroxide cures
Poorer low-temperature set and flex resistance
Higher tension set, less elastic recovery

Molecular weight distribution (MWD) chiefly affects processing properties. Commercial polymers are offered in a wide range of MWD at different Mooney viscosity and ethylene levels. The general effects of narrowing MWD are as follows:

Processing effects:

Increased extrusion rate
Smooth extrusion surface

Lower die swell
Poorer extrusion feed
Greater chance of extrusion melt fracture and edge tearing
Poorer mill handling due to greater adhesion to the mill rolls
Poorer green strength
Poorer calendering
Greater compound flow and tack

Vulcanizate properties:

Increased cure rate
Higher cure state

The level of diene chiefly affects the rate and state of sulfur vulcanization. The effect of increasing diene level on processing and vulcanizate properties is as follows:

Processing effects:

Improved low-pressure, continuous cures
Lower scorch safely
Increased flexibility in selecting accelerators
Shortened shelf life

Vulcanizate properties:

Faster cure rate
Higher modulus
Improved compression set
Decreased elongation
Decreased heat aging resistance

Carbon Black Reinforcement

The most widely used and best reinforcing fillers are the carbon blacks. The important properties are particle size and structures. The medium-particle-size, higher-structure types, such as N550 FEF and N650 GPF, favor reinforcement and increase both hardness and stiffness. The fine-particle-size blacks (for example, N330 HAF) are difficult to mix and disperse and at any given loading produce compounds with much higher viscosity. Coarser or low-structure blacks are suggested when high loadings are required for increased elongation and/or scorch time. The effect of carbon black structure and particle size is summarized in Table 4-6.

Table 4-6 Effect of Carbon Black Structure and Particle Size

		Particle size Fine Coarse	Structure Low High
Processing	Improves	-----→	--→←--
Reinforcement	Increases	←-----	-----→
Optimum loading (for tensile and tear) at increased oil loadings	Increases	-----→	-----→
Mooney viscosity, modulus, and hardness	Increases	←-----	-----→
Mooney scorch, elongation, and resilience	Increases	-----→	←-----
Compound cost	Increases	←-----	←-----

Best overall performance: N550 FEF, N683 APF, N650 GPF-HS
Best for cost compounding: N550 FEF, N650 GPF-HS/N762 SRF-LM blends

The preparation of carbon black-rubber masterbatches is common practice in the industry. Masterbatching consists of adding the appropriate loading of carbon black in the form of a slurry to the rubber prior to coagulation. Table 4-7 provides a summary of the various common grades of blacks used in the synthetic rubber industry.

In plastics applications, medium- and high-color channel blacks are employed for tinting and to provide maximum jetness at low loadings. Carbon black is also used as antiphoto- and antithermal-oxidation agents in polyolefins. As an example, clear polyethylene cable coating is susceptible to crazing or cracking, accompanied by a rapid loss in physical properties and dielectric strength when exposed to sunlight for extended periods of time. The carbon black essentially serves as a black body that absorbs ultraviolet and infrared radiations. It also serves to terminate free-radical chains and hence provides good protection against thermal degradation.

Carbon black is elemental carbon that is differentiated from commercial carbons such as coke and charcoals by the fact that it is in particulate form. Particles are spherical, quasi-graphic in structure, and of colloidal dimensions. Carbon blacks are manufactured either by a partial combustion process or by thermal decomposition of liquid or gaseous hydrocarbons. There are many commercial classifications of carbon black, but all blacks can be generally characterized by their method of production, as follows:

Lampblacks: produced via combustion of petroleum or coal tar residues in open shallow pans (typical sizes 600 to 4,000 Å)
Channel blacks: produced by partial combustion of natural gas or liquid hydrocarbons in retorts or furnaces (400 to 800 Å), coarsest variety (400 to 4,000 Å)
Thermal blacks: produced by the thermal decomposition of natural gas
Acetylene black: produced by the exothermic decomposition of acetylene

Table 4-7 Properties and Rubber Applications of Carbon Blacks

Type (designation)	Average particle size (Å)[a]	Typical Applications
Channel blacks		
Easy processing (EPC)	290	Tire treads, heels, soles of shoes, mechanical goods products
Medium processing (MPC)	250	
Gas furnace blacks		
Semireinforcing (SRF)	800	Tire carcass/bead insulation, footwear and soling, belts,hoses, packings, mechanical goods
High modulus (HMF)	600	Tire carcass/sidewalls, footwear, mechanical goods
Fine furnace (FF)	395	Truck tire carcass, breaker, and cushion
Thermal blacks		
Medium thermal (MT)	4700	Wire insultation/jackets, mechanical goods, belts, hose, packings, stripping mats
Fine thermal (FT)	1800	Natural rubber inner tubes, footwear uppers, mechanical goods, inflations
Oil furnace blacks		
General purpose (GPF)	510	Tire tread base, sidewalls, sealing rings, cable jackets, hose, extruded stripping products
Fast extruding (FEF)	400-450	Tire carcass, tread base, and sidewall, butyl inner tubes, hose, extruded stripping
High abrasion (HAF)	270	Camelback and tire treads, mechanical goods
High abrasion, high structure (HAF-HS)	250	Tire treads, products requiring high abrasion resistance
High abrasion, low structure (HAF-LS)	264	Channel black replacement in natural rubber applications
Intermediate superabrasion (ISAF)	230	Tire treads and camelback, mechanical goods
Intermediate superabrasion, high high structure (ISAF-HS)	225	Tire treads, high abrasion service applications
Intermediate superabrasion, low structure (ISAF-LS)	227	Tire treads, high-abrasion service applications
Superabrasion (SAF)	200	Tire treads, camelback, mechanical goods, heels/soles of shoes
Conductive (CF)	230	Antistatic and conductive rubber products, belts, hose, flooring material

[a] $D_n = \Sigma nd / \Sigma n$, where n = number of particles of diameter d.

The various commercial grades and the specific properties imparted to end-use products via their incorporation depend on particle size (that is, surface area), chemical composition and relative activity of the particle surface, and degree of particle-to-particle association.

Carbon black essentially remains as a discrete phase when dispersed in media such as rubber. The extent of mixture reinforcement imparted to the rubber, or, for example, the intensity of color imparted to a coating formulation, depends primarily on the particle size or surface area.

The definitions used for characterizing average particle size vary. Arithmetic averages are defined as

$$D_n = \frac{\sum nd}{\sum n} \tag{4-1}$$

where n is the number of particles of diameter d.

A more meaningful definition is the surface-average diameter

$$D_A = \frac{\sum nd^3}{\sum nd^2} \tag{4-2}$$

Particle-size distribution of commercial grades can vary from Gaussian to skewed.

Specific surface areas are evaluated by absorption techniques and the arrangement of carbon atoms within a carbon black particle has been extensively studied by X-ray diffraction methods. Carbon black displays two-dimensional crystallinity, which is a structure defined as mesomorphic.

One of the most important properties of carbon black is its surface activity. Carbon itself has proved extremely useful as an adsorbent. All carbon blacks possess a distribution of energy sites, whereby those of the highest energy are first occupied by the adsorbate. As the adsorption process continues, less active sites become filled and the differential heat of adsorption decreases, eventually approaching the heat of liquefaction of the adsorbate at the monolayer coverage. High-energy sites can be progressively reduced by heat treatment. At 3,000°C all high-energy sites are reduced and the substance displays characteristics of a homogeneous surface of uniform activity.

Chain structure is another property of importance, particularly in terms of the properties imparted to rubber. In particular, extrudability, elastic modulus, and electrical conductivity of rubber vulcanizates are sensitive. In general, lampblack and acetylene black have a high degree of structure, whereas channel and low blacks are typically low in chain structure. Most commercial carbon black is available in bead or pellet form. The rubber industry typically employs carbon blacks having an average specific gravity of 1.8.

The relevant properties of carbon black to rubber compounding are particle-size surface area, particle porosity, aggregate structure (that is, bulkiness), the amount of carbon per aggregate, surface activity, and surface chemistry.

Examples of widely used elastomers are SBR, BR, acrylonitrile-butadiene copolymers, and EPDMs. These are examples of polymers in which the molecular chains are not highly oriented when their gum vulcanizates are stretched. Figure 4-5 gives an example of an EPDM, showing the dramatic improvement of tensile and modulus properties with increasing carbon black loadings.

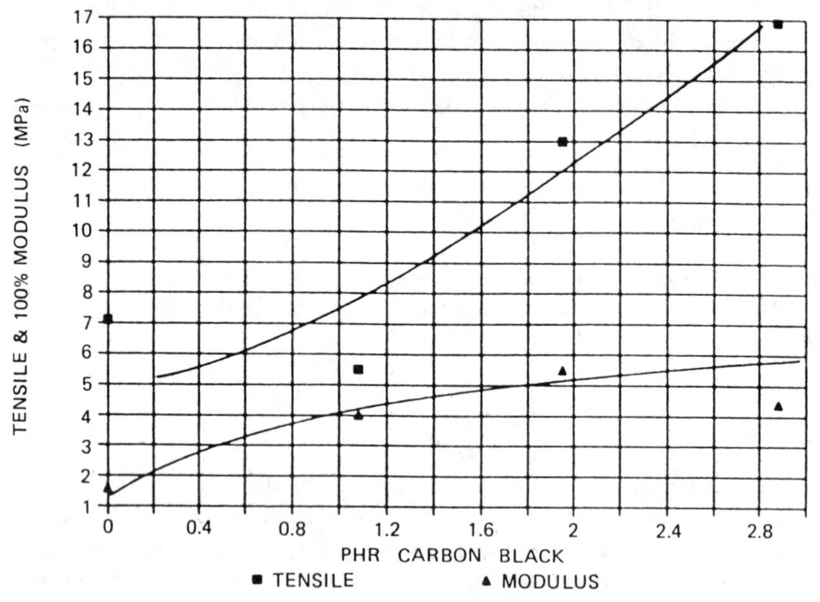

Figure 4-5 Tensile and modulus properties of sulfur-cured EPDM vulcanizate at different carbon black loadings

The preceding example helps to illustrate that the usefulness of elastomers greatly depends on the ability of fillers such as carbon black to impart reinforcing properties. The property of reinforcement is characterized in terms of the vulcanizates' stiffness, modulus, tear strength, rupture energy, cracking resistance, fatigue stress, and abrasion resistance.

It is important to note that some fillers are nonreinforcing, but are instead used as extenders in rubber formulations to reduce raw material costs. These fillers are usually chemically inert and have large particle sizes. Reinforcing fillers are the opposite; that is, typically they are small in particle size and chemically active. Fully reinforcing fillers are those solid additives having mean particle sizes below 50 nm. The two principal additives are carbon blacks and silica. Solid fillers having sizes greater than 50 nm are clays, silicates, calcium carbonates, SFR carbon blacks, and a range of semireinforcing blacks that are usually derived by thermal oxidation.

It is important to make a distinction between the carbon black-rubber interactions of uncured and vulcanized materials. The effects of carbon black on such properties as resilience and stiffness will be different for these two states of the material. In the uncured state the incorporation of carbon black results in viscoelastic changes in the rubber system. In the vulcanized state, carbon black incorporation causes viscoelastic property changes which directly alter the rubber network. The specific properties affected in the vulcanizate are the modulus, dynamic properties, and hysteresis.

Commercial carbon blacks used in rubber compounding applications are usually classified in terms of their morphology (that is, particle-size surface area, vehicle absorptive capacity). In the ASTM nomenclature system for carbon black, which is strictly used in the United States, the first digit is based on the mean particle diameter as measured with an electron microscope. Other commonly used methods for classifying blacks by size are iodine number, nitrogen adsorption, and tinting strength. Important also are the shapes of the individual particles and of the aggregates. DBP absorption is the principal technique employed for measuring the irregularity of the primary aggregates.

The term *structure* characterizes the bulkiness of individual aggregates. Often when carbon black is mixed into rubber, some of the aggregates fracture. In addition to fracturing, the separation of physically attracted or compacted aggregates (microagglomerates) can occur. Aggregates can be held together fairly rigidly because of the irregular nature of their structures. Microagglomerates can also be formed during mixing. That is, high-shear forces can cause microcompaction effects wherein several aggregates can be pressed together. Excessive microagglomeration can lead to persistent black network structures which can affect extrusion and ultimate vulcanizate properties.

The relationship of carbon black morphology to the failure properties of rubber vulcanizates has been studied extensively. Strength properties are usually enhanced with increasing black surface area and loading (references at the end of the chapter provide detailed discussions). The upper limit of black loading for maximum tensile strength and tear resistance depends on carbon black fineness and structure, with the former usually having the greater effect. Coarser, lower-structure blacks generally show peak-strength properties at higher loadings. In terms of the ultimate level of strength reinforcement for different blacks, structure is significant as a dispersing aid, which may be attributable to better bonding between black and polymer. High structure is also important at lower black loadings, particularly for tear resistance.

There is evidence that the strength-reinforcing properties of fillers are directly related to modulus development. These properties are derived from the large stresses held by the highly extended polymer chains attached to the immobile particles. Boonstra [6] has related high tensile strength to energy dissipation by a slippage mechanism at the filler surface. This is supported by the fact that high-surface-area, inactive (partially graphitized) carbon blacks give very high tensile strength under standard testing conditions. Under more severe conditions, however, adhesion between black and polymer becomes important.

Andrews and Walsh [1] have shown that the path of rupture through a filled vulcanizate passes from one filler aggregate to another, which are sites of high-stress concentration. One approach to increasing strength is to lengthen the overall rupture path. The finer the filler and the higher the loading, the greater the effective increase in total cross section. There is a limiting point, however, when the packing of the filler aggregates becomes critical and they are no longer completely separated by the polymer. Preferential failure paths between aggregates may then be formed. These, combined with the fact that a smaller amount of polymer is being strained, can result in failure. The level of filler-polymer adhesion is important, especially as the upper limits of loading (aggregate packing) are reached.

Filler-polymer adhesion is important to the amount of elastic energy released by internal failure. Strongly adhering fillers enlarge the volume of rubber that must be highly strained during the process of rupture. It is worth noting that the relative importance of different carbon black parameters that influence vulcanizate properties depends greatly on the conditions of testing.

Carbon black fineness is the dominant factor governing vulcanizate strength properties. Tensile strength, tear resistance, and abrasion resistance all increase with decreasing black particle size. Cyclic processes such as cracking and fatigue are more complex because they may involve internal polymer degradation caused by heat build-up. Beatty [4] points out that it is important to note whether or not the end-use application represents constant energy input or constant amplitude vibrations. The latter mode of service greatly reduces the fatigue life of rubber compounds containing fine or high-structure blacks. Beatty has reported no significant particle size effect on fatigue on the basis of equal energy input.

The relationship of carbon black fineness to failure properties is not straightforward, since it is difficult to separate the effects of aggregate size and surface area. It is likely that both are important, since aggregate size relates to the manner in which the surface area is distributed. In general, decreasing aggregate size and increasing black loading reduce the average interaggregate spacing, thereby lowering the mean free rupture path. Increasing black fineness and loading can be viewed in terms of either increased interface between black and polymer or increased total black cross section (lower aggregate spacing). Both show a strong effect on tensile strength up to the maximum tensile value that can be achieved.

The ultimate value for tensile strength across all blacks appears to be determined by either aggregate size or specific surface area. Tensile strength can be improved to a limiting value by increasing the loading. It is not possible, however, to match the ultimate tensile of a fine black by increasing the loading of a coarse one, at least not without other modifications (for example, improving the ultimate dispersion or the bonding between black and polymer).

Blends of two or more elastomers are used in a variety of rubber products. The compatibility of elastomers in terms of their relative miscibility and response to different fillers and curing systems is of great importance to the rubber compounder. From the standpoint of carbon black reinforcement, certain combinations of polymers can give less than optimum performance if the black is not proportioned properly between them. Note that an equivalent volume proportionality of the black is not always desirable. This depends on the nature of the polymers and their relative filler requirements in terms of strength reinforcement.

The compatibility of elastomer blends has been studied by a variety of different techniques, including optical and electron microscopy, differential thermal analysis (DTA), GPC, solubility, thermal and thermochemical analysis, and X-ray analysis. These and other methods have been discussed by Corish and Powell [15].

Microscopic methods are especially useful in studying the phase separation (zone size) of different polymer combinations and in determining the relative amounts of filler in each blend component. Walters and Keyte [53] studied elastomer blends by means of phase contrast optical microscopy, where the contrast mechanism is based on differences in the refractive indexes of the polymers. Their work covered blends that included NR/SBR, NR/BR, and SBR/BR. Their results indicated that few, if any, elastomers can be blended on a molecular scale. They also provided direct evidence that fillers and curing agents do not necessarily distribute proportionately.

The term *compatibility* can be defined in terms of the glass-transition temperature (T_g). Blends that have T_g values of the individual polymer components are considered incompatible, while those that result in a single, intermediate T_g can be viewed as compatible. Also, solubility parameters are considered a prerequisite to compatibility. In the category of compatible blend combinations Corish [14, 15] lists NR/SBR, NR/BR, and SBR/BR. Incompatible blends are NR/NBR, NBR/BR, and NR/CR.

The compatibility of elastomer blends can also be defined from the standpoint of interfacial bonding between the various polymer phases. The technique of differential swelling, using solvent systems and temperatures such that one elastomer is highly swollen and the other is below its temperature with the polymer chains remaining tightly coiled, can be used. Thus, the polymer below its temperature could be treated in the same manner as a filler with respect to the way it restricts the swell of the other polymer. A high degree of swelling restriction is indicative of interfacial bonding between phases, while the reverse is analogous to the dewetting that occurs with nonreinforcing mineral fillers in elastomer networks. Also, the type of curing system is important in achieving interfacial bonding.

The importance of the proper curing system in blends of low-functionality rubber with more unsaturated or polar elastomers must also be emphasized. Large amounts of polymer can be extracted from blends when unsatisfactory curing systems (for example, sulfur-ZnO alone) are employed. This can be attributed to curative diffusion because it occurs whether or not the curatives are added to the blend or to the low-functionality rubber alone.

Good blending is favored by a similarity of solubility parameters and viscosity. For polymers varying in unsaturation or polarity, cure compatibility must be considered in terms of both properly cross-linking each polymer component and achieving satisfactory interfacial bonding. In contrast, the addition of fillers to elastomer blends can significantly alter the state of the polymer phases. Where large differences in unsaturation exist between the polymers in a blend, staining methods may be used to render the high-unsaturation polymer more opaque to provide contrast for either light- or electron-microscope analysis.

To the point, the contributing aspects of reinforcement are dispersion, aggregate breakdown and interactions, and distribution between the separate phases of a polymer blend. For blends of dissimilar elastomers (for example, in unsaturation, polarity, crystallization, or viscosity), problems can arise in achieving optimum black distribution in the end product. When fillers are added to elastomer blends, one very obvious change is a reduction in the size of the separate polymer zones. In other words, there is a preferential location of the black in one of the blend components. This phenomenon may be related to both molecular structure and relative viscosity. The great mobility and possibly more linear structure of one of the blend component molecules apparently enables better wetting of the black surface. This characteristic is often observed in SBR/BR blends. The lower initial viscosity of the BR would also tend to favor the acceptance of the black. It is, however, possible to alter the normal pattern of distribution in terms of how the filler is added, along with mixing conditions. Other polymer factors that favor preferential filler adsorption in elastomer preblends are higher unsaturation and polarity.

Another phenomenon contributing to reinforcement is carbon black transfer. This is the ability of a carbon black to migrate from one polymer to another during mixing. Transfer is favored by low-heat history and high extender-oil content. Both these parameters minimize interaction between the black and polymer. Hence, solution or latex masterbatches tend to favor transfer, whereas hot mixing in a Banbury does not. Black transfer also occurs when a masterbatch of a low-unsaturation rubber is cut back with a high-unsaturation polymer. Polarity is also a factor.

As noted earlier, carbon black is widely used as a reinforcing filler; however, the unique property of improving tear and abrasion resistance in rubber vulcanizates is not fully understood. It is likely that carbon black surface and rubber interact both chemically and by physical adsorption. The parameters important to this interaction are capacity, intensity, and geometry. The total interface between polymer and filler is expressed in units of square meter per cubic centimeter of vulcanizate or compound. The intensity of interaction is determined by the specific surface activity per unit area. It should be noted that adsorptive energies vary greatly in different locations on the black surface and it is likely that this distribution is responsible for the variability in properties among different types of carbon blacks. Geometrical properties are characterized by the structure, which is basically anisometric, and by particle shape and porosity. Greater anisometry results in looser particle packing. The void volume is used as a measure of the packing density.

Carbon black particles form irregular structures which tend to break down during intensive mixing. It is therefore a combination of carbon black reactivity; rubber and black chemical, physical, and rheological properties; and the conditions of mixing which establish the final strength properties of the vulcanizate. As a rule of thumb, high-structure blacks (for example, ISAF) impart a higher modulus (at 300 percent elongation) than that of a corresponding normal black. The high modulus is determined by both the anisometry and surface activity. The separate influence of these properties can be observed by heat treating these blacks so that their properties approach that of graphite. Through recrystallization, highly active sites on the carbon black surface lose their high activity. In this situation the entire surface area becomes homogeneous and adsorption energies approach their lowest state. Figure 4-6 illustrates the effect of this reduction in surface activity on vulcanizate properties. Heat treatment typically results in a minor decrease in surface area and further reductions in tear resistance. Modulus and elongation can be decreased by a factor of 3 or 4. The decrease in properties can be attributed to the removal of highly active sites from the surface. Boonstra [5] shows that this is accompanied by a decrease in water adsorption and propane adsorption.

Microscopic examinations of thin sections of vulcanizate in which carbon black or other fillers and polymer were mixed for a short time only reveal that this additive exists as coarse agglomerates in an almost pure rubber matrix, without

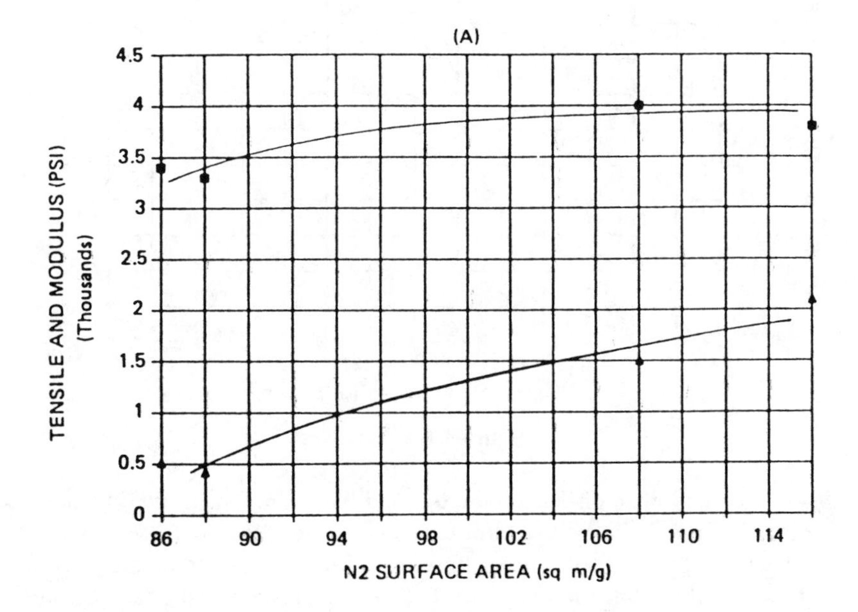

Figure 4-6 Effect of graphitization of black on vulcanizate properties: (A) physical properties; (B) vulcanizate elongation; (C) scorch

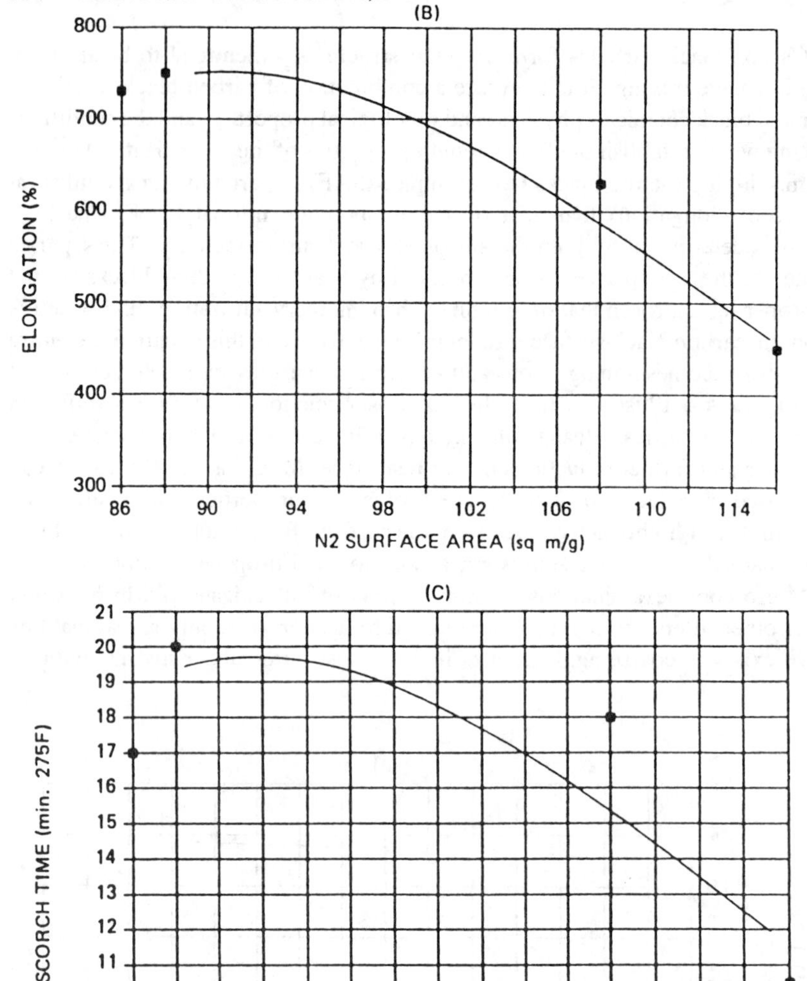

Figure 4-6 ***(Continued)***

almost any colloidal dispersion of the additive. Only after mixing has been continued at greater intensity and/or for longer periods can one observe how these agglomerates gradually disappear, making room for an increasing amount of additive. In all cases the initial product of mixing is comprised of an agglomerate that is formed by the penetration of rubber in the voids between carbon black particles or other additives, such as mineral fillers (for example, clays) under the pressure and shear that builds up during Banbury mixing and on a roll mill. Once these voids are filled, the additive is incorporated but not yet dispersed. As soon as the agglomerates are formed, continuous mixing exposes them to shearing forces which break them up and eventually disperse them. In other words, both agglomerate formation and breakdown take place almost simultaneously.

Mineral Filler Reinforcement

Inorganic fillers are used in white or colored compounds or as extenders in carbon black stocks to reduce cost. Silicas, clays, talcs, and whitings are examples of the classes of mineral fillers most commonly used. Frequently, a combination of fillers is used to obtain the required balance of reinforcing properties, processability, and economics for the particular application. Fine-particle-size silicas are the most reinforcing of the mineral fillers. However, they are difficult to disperse, retard cure, and produce boardy, hard stocks. Hard clays and talcs are easier to mix, provide some reinforcement, and retard cure to a lesser extent than silicas. To overcome the retarding effect, small amounts of diethylene glycol or polyethylene glycols are commonly added.

Talcs, whitings, and soft clays impart little reinforcement but aid processing by reducing compound viscosity. Table 4-8 provides guidelines on the effects of different mineral fillers.

Table 4-8 Effect of Mineral Fillers on Compound Properties

Compound	
Good reinforcement	Silica, hard clay, platy talc
Good mill mixing	Silicate, calcined clay, platy talc
Good extrusion	Silica, platy talc, calcinated clay
Good internal mixing	All except ground whiting
Lowest viscosity	Platy talc, whiting
Vulcanizate	
Highest hardness	Silica, silicate
Highest modulus	Silica, platy talc
Highest tear	Silica, silicate
Good compression set	Whiting, soft clay
Good water resitance	Calcined clay, platy talc, soft clay

Plasticizers

Ethylene-propylene rubber is compatible with paraffinic or naphthenic process oils. Aromatic oils should be avoided because they have a detrimental effect on cure. Viscosity is an important consideration when selecting an oil plasticizer. In general, the higher-viscosity oils enhance physical property development, improve heat resistance, and minimize shrinkage. Lower-viscosity oils improve both resilience and low-temperature flexibility, and tend to reduce compound viscosity more but are more fugitive.

Because of the very high filler loadings often employed in EPDM compounds, the oil performs several important functions:

Improves wetting and incorporation of fillers
Reduces power consumption of the mix equipment
Lowers batch temperature
Reduces the risk of compound scorch
Improves extrusion and all other shaping operations

As an extender, oil almost always lowers compound cost. Table 4-9 provides a guide to the selection of process-oil type.

Table 4-9 Process-Oil Types Typically Used

For:	Use:
Best physical properties	High-viscosity oils
Maximum heat resistance	High-viscosity paraffinic oils
Highest resilience and best low-temperature flex	Low-viscosity oils
Lowest compound viscosity	Low-viscosity oils
Peroxide cures	Only paraffinic oils

Use of Contour Maps

A contour map is a useful method of presenting rubber compounding data. It is similar to a line of constant elevation on a topographical map, representing a single value over the range of the grid coordinates. In rubber technology the coordinates are usually expressed in loading levels of fillers and plasticizers in the compound.

The family of curves so generated is called a contour plot and can be used in two different ways.

1. To determine the general effect of the compounding variables on a property of interest
2. To select a combination of filler and plasticizer loading to produce a desired property value

A typical contour map for a polymer is shown in Figure 4-7. Rubber manufacturers will typically supply these processing maps for different grades of their polymers.

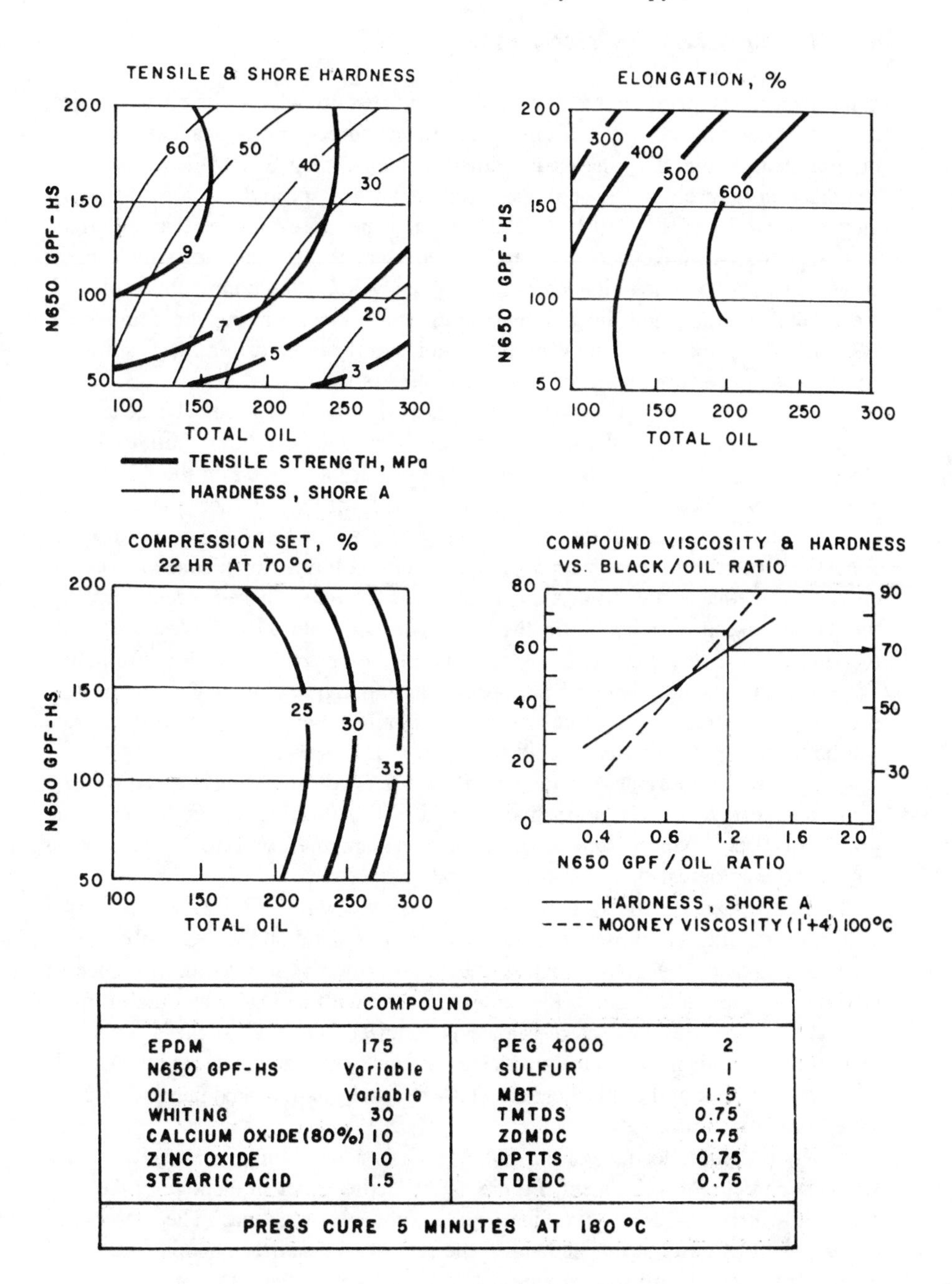

COMPOUND			
EPDM	175	PEG 4000	2
N650 GPF-HS	Variable	SULFUR	1
OIL	Variable	MBT	1.5
WHITING	30	TMTDS	0.75
CALCIUM OXIDE (80%)	10	ZDMDC	0.75
ZINC OXIDE	10	DPTTS	0.75
STEARIC ACID	1.5	TDEDC	0.75
PRESS CURE 5 MINUTES AT 180 °C			

Figure 4-7 Construction and use of typical contour map for a particular polymer prepared in a sulfur cure compound

4.4 VULCANIZATION PRINCIPLES

Vulcanization is primarily a cross-linking process in which chemical bridges, characteristic of the cure system employed, form a three-dimensional network with the polymer chains. The chemical method of vulcanizing EPDM depends on the chemical structure of the polymer. Ethylene and propylene are the basic monomeric units of EPM/EPDM. The structures are more complex than described earlier in that we have not attempted to represent the probable distributions of ethylene and propylene units determined by the ratio of the monomers employed, the catalyst system, and the polymerization conditions. Among the commercial EPM/EPDM producers, many products are significantly different, despite the commonality of Mooney viscosity or monomer ratio.

EPM copolymer is cross-linked exclusively via a free-radical mechanism, usually accomplished by the decomposition of organic peroxides; initiation with electron beam radiation is also possible. The tertiary hydrogen on the EPM main chain is abstracted by a peroxy radical to form a polymer radical. This subsequently combines with adjacent radicals to complete the cross-link. In general, free-radical vulcanization is enhanced by polyfunctional coagents such as ethylene dimethacrylate, triallyl cyanurate, and the like. These reagents increase the cross-linking efficiency of the peroxide and minimize undesirable side reactions. However, the decomposition rate of the peroxide, which is a function of temperature, is not altered. Acidic compounding ingredients such as stearic acid promote ionic breakdown of peroxides and should be avoided. Calcium stearate can be used in place of stearic acid with peroxide cures.

One undesirable side reaction with free-radical vulcanization is the competing chain breakdown at the tertiary hydrogen site. For this reason, higher ethylene grades of EPM, which have fewer such sites, are preferred for vulcanization whenever rheological or end-use considerations permit.

The peroxide cure system can also be used with EPDM; the principal vulcanization site is at or near the side-chain double bond on the diene. Reactivity rates are significantly greater than for EPM copolymers due to the presence of allylic hydrogen on the diene. Peroxides are used with EPDM when the ultimate in heat and compression set resistance is required. Generally, these are high-quality compounds in which the plasticizer level is minimized to avoid interference with the cure. If a higher plasticizer level is necessary, additional peroxide can be added to compensate.

The peroxide decomposition rate, which is a function of temperature, is a key factor in cure development. Figure 4-8 is a cure meter plot showing the cure rate with three different peroxides on a rubber compound. The onset and development of cure are dependent on the rate of free-radical generation as well as the structure of a given peroxide.

Convenient sulfur vulcanization through the diene is used in the majority of EPDM applications. The type of cross-link is related to the specific cure system used. High-sulfur systems favor the formation of a polysulfidic bond. These

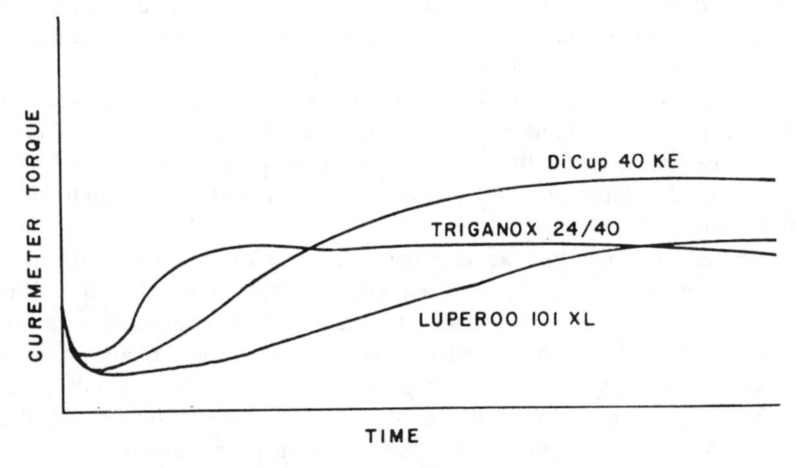

Figure 4-8 Effect of peroxide on cure for an EPDM

vulcanizates offer high stress-strain properties and good flexibility, but thermal stability is marginal compared to systems generating monosulfide bonds.

Vulcanization rate is related to the amount and type of diene present in the polymer, as well as to the cure system. When properly boosted with ultra accelerators, a nominal 5 wt percent ENB grade can be vulcanized in fast cure cycles. However, the amount of diene is not the only controlling factor: narrow-MWD polymers attain a higher cure rate and state than do broad-MWD polymers. This is illustrated in Figure 4-9. The basic sulfur cure systems for EPDM consist of:

Cross-link agent: sulfur, sulfur donor
Primary accelerator: thiazole
Ultra accelerator: thiurum, dithiocarbomate

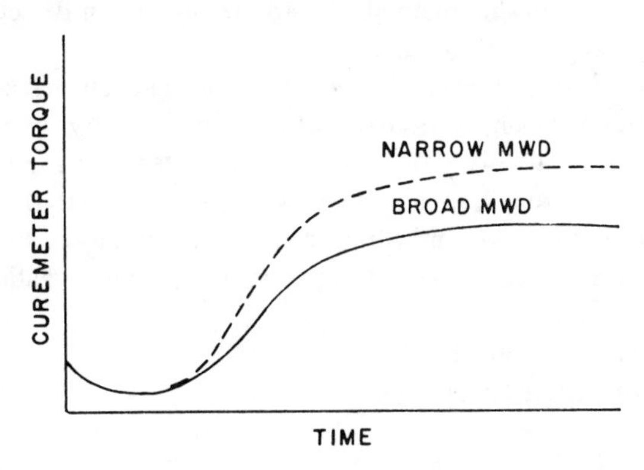

Figure 4-9 Effect of polymer MWD on cure state

Compared to more highly unsaturated rubbers, very large amounts of thiuram and dithiocarbamates are used in EPDM; up to 2 to 3 parts per hundred rubber (phr) of each can be used in some applications.

Several general and special-purpose sulfur cure systems are summarized in Table 4-10. Principal variations are in cost, cure rate, processing safety, and type of sulfidic cross-link. These sulfur cure systems are intended to be starting points for compound development; they frequently are modified for individual applications and customer preferences.

Bloom resistance is also a factor in the choice of a sulfur system. Bloom is a gradual migration to the surface of a partly insoluble by-product of sulfur vulcanization, usually a zinc dialkyl dithiocarbamate. Bloom is especially sensitive in press cures; limited amounts of mixed accelerators give less bloom than do equivalent amounts of a single accelerator. Polyethylene glycol (for example, PEG 3350) inhibits the formation of bloom by increasing the solubility of the accelerators. Some sulfur cure systems that bloom in press cures do not give bloom in steam cures.

A variety of methods are used to achieve vulcanization. In compression molding the uncured stock is placed directly in the mold cavity. The mold is then assembled and placed in the hydraulic press. The press closes the mold and causes the stock to fill the cavity. When cure is effected, the mold is removed from the press, disassembled, and the rubber parts removed.

Transfer molding is a refinement of the compression molding process, in which the mold is tightly closed before the cavity is filled. Stock is "transferred" from a holding pot to the cavity under pressure.

Injection molding is similar to transfer molding except that the stock is metered into the mold cavities through runners fed by an injection head. On some machines one injection head may serve several presses and molds. This higher degree of automation permits short, high-temperature cures with short change times.

Open steam in an autoclave has long been used to cure extruded and calendered products. With this method, the article may be in direct contact with the steam or wrapped with fabric tape.

Continuous vulcanization combines processing and curing steps into one continuous operation on many extruded profiles. Traditionally, extruded profiles and hoses have been separately processed and vulcanized. During the processing step, the profile is extruded in a continuous operation, but the extrudate is accumulated for later batch vulcanization, in a steam autoclave, for example.

Basically, there are three stages to the continuous vulcanization process:

1. Formation of the profile
2. Heating to curing temperature
3. Curing

Table 4-10 Typical Sulfur Cure Packages[a]

	Type of sulfidic crosslinks							
	Polysulfidic, C—S_x—C							Monosulfidic, C—S_1—S_2—C
	Cure system A: General purpose, low cost	Cure system B: Fast, nonblooming	Cure system B_1: Fast, nonblooming	Cure system C: Fast, general purpose	Cure system C_1: General purpose, good balance of properties	Cure system D: Low cost, nonblooming	Cure system D_1: Low cost	Cure system E: Special purpose heat and compression set
	1.5 Sulfur 0.5 MBT 1.5 TMTDS	2.0 Sulfur 1.5 MBT 0.8 TMTDS 0.8 TDEDC (80%) 0.8 DPTTS	2.0 Sulfur 2.5 MBT 0.8 TMTDS 0.8 TDEDC 1.0 DPTTS 1.0 ZDEDC 0.8 ZDBDC	1.5 Sulfur 0.5 MBT 3.0 TMTDS	1.5 Sulfur 0.5 CUPS 3.0 TMTMS	1.5 Sulfur 2.0 MBTS 0.8 TMTDS 0.8 ZDEDC	1.5 Sulfur 1.5 CBS 0.8 ZDBDC 0.8 TMTDS	0.5 Sulfur 3.0 TMTDS 2.0 DTDM 3.0 ZDMDC 3.0 ZDBDC
	Medium cure rate Good heat resistance Fair compression set Good scorch resistance Nonblooming in steam Low cost	Fast cure rate High tensile strength Fair compression set Nonblooming in press and steam	Fast cure rate High tensile strength Fair compression set Nonblooming in press and steam	Fast cure rate High tensile strength Good heat resistance Good compression set Good scorch resistance Nonblooming in steam Low cost	Good overall balance of properties, including scorch safety stress–strain and compression set	Medium cure rate Good scorch resistance Nonblooming in press and steam Low cost Fair compression set	Fast cure rate Good scorch resistance Good CV cure Low cost Fair compression set	Medium cure rate Best (nonperoxide) heat aging Best (nonperoxide) compression set Excellent scorch resistance Slight bloom in steam
Press cure, 20 min at 160°C								
Hardness, Shore A	70	73		71	71	72		70
100% modulus, MPa	3.7	4.8		4.3	4.0	3.7		3.7
300% modulus, MPa	10.7			11.9	10.9	10.6		10.6
Tensile strength, MPa	12.3	12.8		12.3	11.9	12.1		11.7
Elongation, %	400	280		330	380	390		390
Compression set B, (press cure 25 min at 160°C) 22 h at 100°C, %	38	39		29	23	40		18
Mooney scorch, MS, at 132°C, t_3, min	13	7		11	18	13		12

[a] Base compound: Vistalon TR 5600 (Exxon Chemical Co.), 100; N550 FEF, 50; N762 SRF, 150. Process oil type 103 (Flexon 800), 120; zinc oxide, 5; stearic acid, 1.

Heat may generally be inside the extrudate by dielectric heating (ultrahigh frequency, UHF), by friction (shear head), or by heat transfer from the outside [hot air, fluid bed, liquid curing media (LCM)]. Curing is achieved by maintaining the temperature for the time needed to cross-link the rubber fully.

Typically, some of these operations may be combined as shown in Figure 4-10. For example, shear head or UHF heating may be followed by one or more hot-air curing stages. Shear-head extrusion followed by one-stage hot-air curing is an efficient method of continuous vulcanization.

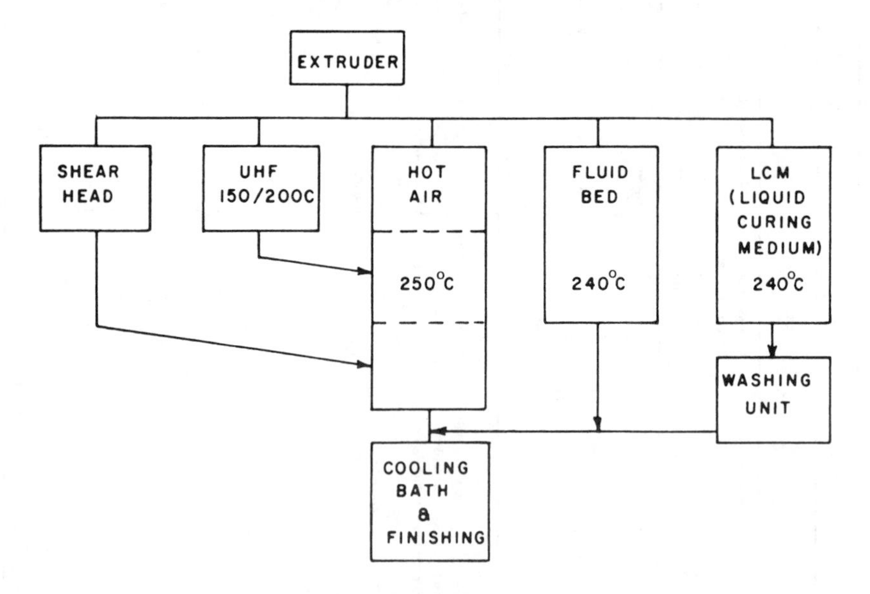

Figure 4-10 Continuous vulcanization process

In a shear head, the compound can be heated from 150° to 220°C. Extrudability can be good, even for somewhat complex profiles, since the compound viscosity decreases with shear rate. Narrow-MWD EPDM grades give higher heat build-up because of their lower dependence of viscosity on shear rate.

4.5 PROCESSING OPERATIONS

The general effects of major variations in carbon black and oil content are illustrated by the processing map in Figure 4-11. Intrinsic properties of the polymer determine the shape and location of the *optimum processing zone* for a particular polymer. A typical processing map for a polymer compounded EPDM is shown in Figure 4-12. The boundaries are not strict limits, only indications of usual performance. Within the optimum processing envelope, simple compounds usually process well. Outside the envelope more compound development is frequently required for acceptable processability.

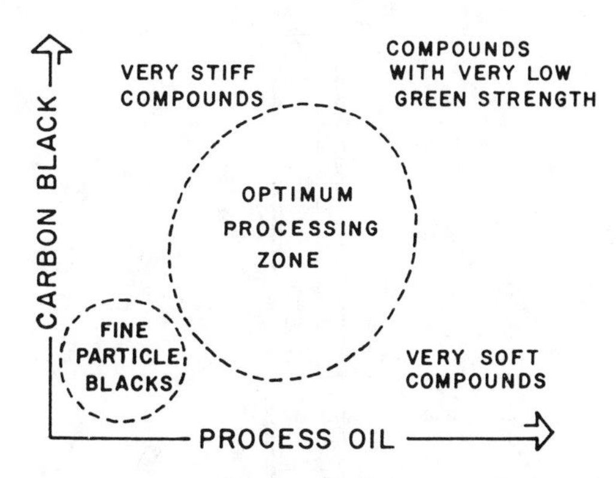

Figure 4-11 Features of a processing map

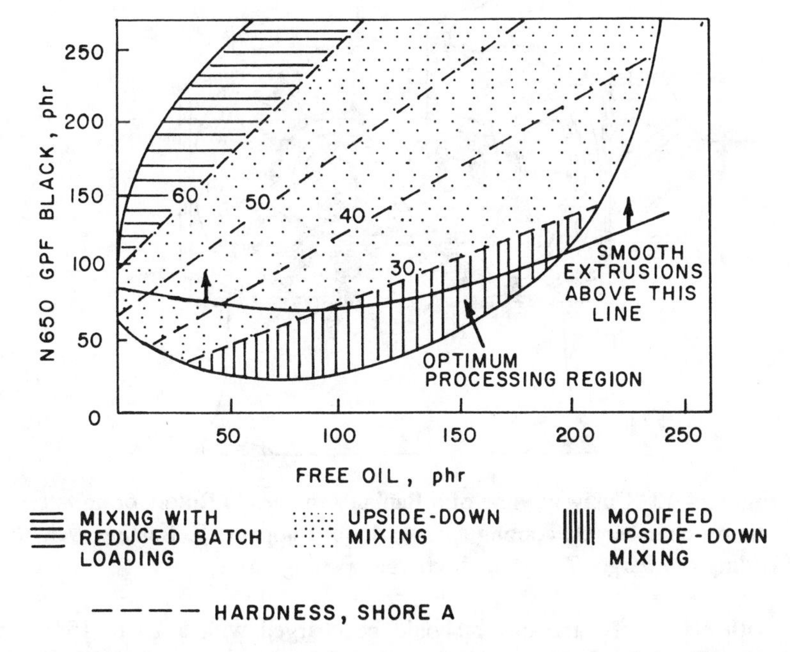

Figure 4-12 Typical processing map for a compounded polymer

In the following paragraphs, the various processing operations used for compounding and forming rubber articles are described qualitatively. Detailed treatments of each of these process operations are given in subsequent chapters.

The first operation of importance is mixing. The prime objective in mixing is to obtain a homogeneous dispersion of carbon black oil and polymer (that is, a compound). There are various types of mixer configurations employed throughout the rubber industry. Cheremisinoff [11] describes several of these. Figure 4-13 illustrates the most common internal type mixer, called a Banbury.

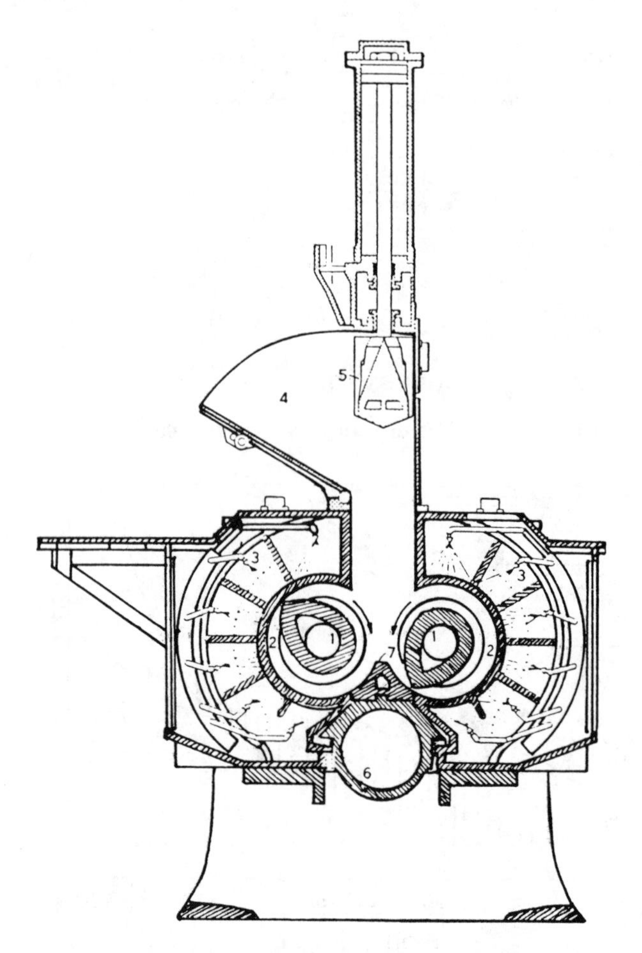

Figure 4-13 Cutaway view of a Banbury mixer. 1, Rotors or agitators; 2, mixing chamber; 3, cooling sprays; 4, feed hopper; 5, floating weight; 6, sliding discharge; 7, saddle discharge opening

With elastomers, the mixer should be charged with a 10 to 15 percent overload for the most rapid and efficient mixing. Most compounds having a cured hardness of 40 to 80 Shore A are thoroughly mixed after 3 or 4 min in a large internal mixer at low speed (20 rpm). For high loadings, the "upside-down" addition of ingredients to the mixer is generally most effective. This involves the addition of all the fillers and oil prior to the addition of polymer. However, soft sponge stocks, compounds with very high oil levels, and low viscosity mineral-filled stocks should be mixed conventionally or "right side up."

Dump temperatures in the range 125° to 150°C are suggested for the N762 SAF and N550 FEF loaded stocks. Accelerated compounds or those containing blowing agents should be dumped at temperatures below 110°C. Highly loaded, high-hardness, high Mooney compounds should be mixed with no overload and include 10 phr microcrystalline wax for the best dispersion. Often, these tough compounds require two-pass internal mixing to obtain optimum vulcanizate properties. Suggested internal mixing cycles are shown in Table 4-11.

Table 4-11 Suggested Internal Mixer Procedures (One-Pass Mixing)

Time (min)	Addition
	Right side up
0	Elastomer, zinc oxide, stearic acid
0.5	One-half filler
1.5	One-fourth filler, one half-oil
2.5	One-fourth filler, one-half oil
3.5	Sweep, curatives
4.0	Dump below 125°C
	Upside down
0	Filler, oil, zinc oxide, stearic acid
0.5	Elastomer
2.5	Sweep, curatives
3.0	Dump below 125°C
	Modified upside down (for very low hardness)
0	One-half filler, two-thirds oil, zinc oxide, stearic acid, four-tenths filler
0.5	Polymer, ram down
4.5	At 140°C, remaining filler, one-third oil
≈10	Dump at 140°C

The term *incorporation* refers to the wetting of carbon black with rubber. During this operation, entrapped air is squeezed out between the voids of the rubber and carbon black particles. Early stages of mixing reveal that as the carbon black becomes incorporated, relatively large agglomerates (on the order of 10 to 100 μm) form. Cotten [16, 18] suggests that the time required for full carbon black incorporation can be determined by measuring the time required to reach the second power peak during a mixing cycle. Cotten [18] performed mixing studies in a Banbury mixer, observing from microscopic examination of rubber-carbon black compounds the progression of black dispersion at different times. The rates of carbon black dispersion in this study were computed from maximum torque data. A typical power curve generated by the author in reproducing Cotten's work is shown in Figure 4-14 using an oil-extended EPDM. Following Cotten, the rubber was first masticated for about 2 min and then the rotors were stopped. Carbon black was charged in the chute, the mixer ram inserted, and the rotors started again. Mixing times were measured from the instant when the rotors were restarted. The carbon black incorporation time was taken to be the time required to attain the second power peak shown in Figure 4-14.

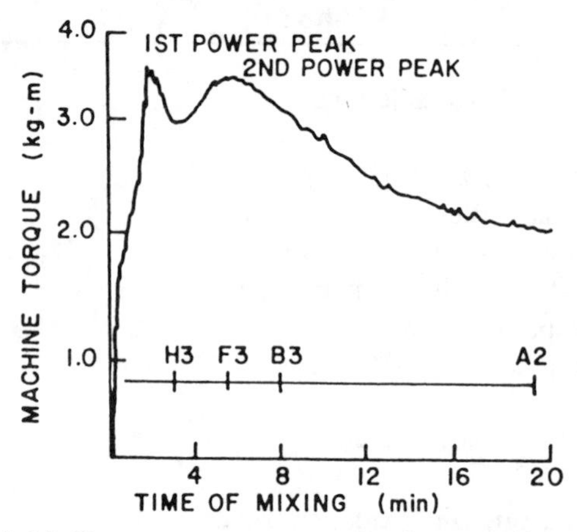

Figure 4-14 Power curve generated using an oil-extended EPDM

Bound rubber in studies such as this can be measured by standard solvent extraction techniques. For example, toluene at room temperature or boiling heptane or hexane extractions can be used with a known aliquot of rubber sample. From a measurement of the residue weight, one can calculate the percentage of insoluble polymer.

The percentage (by volume) of unincorporated (nonwetted) carbon black can be estimated from density measurements. The volume of air in the batch (V) at time t can be computed from the difference in densities (ρ' and ρ) at time t and the time at the termination of mixing when carbon black dispersion is better. The formula used for this calculation is

$$V = \frac{B}{\rho'} - \frac{B}{\rho} \tag{4-3}$$

where B is the batch weight (phr).

The first power peak corresponds to the ingestion of the batch into the mixing chamber. This coincides with the instant that the ram in the loading chute reaches the bottom of its flight and thus removes any additional hydrostatic pressure from the mixing chamber. Cotten has found that the fraction of undispersed carbon black decreases linearly with the time of mixing at the point where the second power peak occurs. At shorter mixing times the compound can appear inhomogeneous and crumbly. The mixing times can be observed to be inversely proportional to the rotor speed. When normalized, data tend to collapse to a single linear correlation. Cotten's work shows that there is a strong correlation between the incorporation time and the rate of incorporation as computed from the slope of the regression lines for a change of densities with time for carbon black compounds in a single polymer.

The existence of a double power peak depends in part on the properties of the carbon black itself. Using pelletized carbon black will result in an absence of the first power peak because no additional hydrostatic pressure is applied when the mixer chute is lowered (hence, the measured torque is no longer affected). The time to reach the second power peak, however, remains the same. Cotten notes that with fluffy carbon black, the time to reach the second power peak increases if no pressure is applied. Large agglomerates of carbon black form during the initial mixing stages regardless of the type of carbon black used.

Other studies by Turetsky et al. [51] and Smith and Kasten [45] have shown good correlation between the measured black incorporation time (BIT) and weight-average molecular weight, M_w, of polymers. Incorporation times will increase sharply with increasing polymer molecular weight. This effect tends to mask expected decreases in incorporation times when the Mooney viscosity is decreased through the addition of oil.

Decreasing concentrations of carbon black and increasing oil loading tend to reduce incorporation time. This is generally thought to be related to a lowering of the polymer viscosity.

Typical filler-reinforced elastomers are contained in the rubber filler, curatives, plasticizers, and stabilizers. After mixing and curing a masterbatch, both physical and chemical cross-linking processes transform the system into a network structure. The structure is composed essentially of two networks: a stationary and a transient network. The stationary network is comprised of chemical cross-links that are formed by both curatives and fillers. This network also includes permanent crystalline structures from crystallizable polymers. In contrast, the transient network is comprised of trapped entanglements and transient ordered structures. Payne and Watson [39] and Lee [32] have proposed two mechanisms that explain the reinforcing effects of filler in elastomers. The first of these mechanisms can be described as a hydrodynamic interaction in which the filler particles are responsible for the reinforcement. The second mechanism is that of strong chemical interactions between the filler and the matrix. Both of these reinforcing interactions contribute to the mechanical properties of filler-reinforced polymers. As noted earlier, the mechanical properties of engineering importance are the tensile modulus, the tensile strength, and the ultimate stretch ratio. Tensile modulus is mathematically defined as the slope of the stress-strain curve at zero strain for a given masterbatch. Tensile strength is defined as the force at break per unit area of the original sample. The ultimate stretch ratio is the uniaxially fractured gauge length divided by the original gauge length of the filler-reinforced polymer specimen.

The milling of properly compounded rubber stocks can easily be accomplished on cool or warm roll mills. Hot stock freshly dumped from the internal mixer on cool slabs will band immediately and stay on the cooler roll. Sheet-off operations are fast and trouble free. Mill-roll sticking of soft stocks can be avoided by warming the mill rolls slightly. Figure 4-15 illustrates the operation of milling.

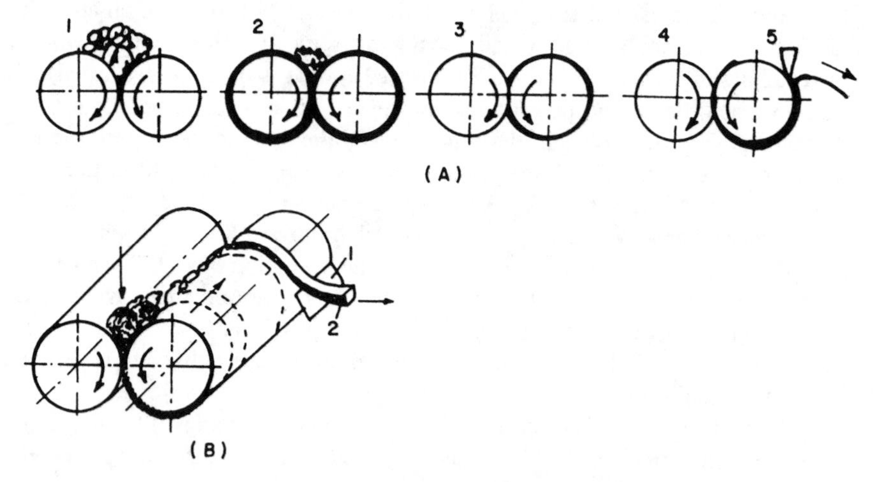

Figure 4-15 Milling and mill mixing

Low-to-intermediate Mooney and oil-extended polymers can be mixed on an open mill. The polymer has an initial tendency to resist banding, which is overcome after a few passes through a tight nip. Once a band is formed, the stearic acid, zinc oxide, and filler should be added as rapidly as the polymer will incorporate these ingredients, without concern about large amounts of unmixed filler on top of the mill. Process oil should be added with the carbon black to avoid mill surface lubrication and loss of bondage.

Often to save time, a manually prepared premix of black and oil may be made. Such a premix consists of all the oil plus sufficient black to make a sort of dry paste—usually about equal weights of black and oil. This premix may not be necessary if a mill apron is used. Leaving a small amount of stock on the mill, or *seeding* the next batch to be mixed, can also save mixing time. During mixing the batch may flip to the faster roll. Since a temperature differential is maintained between the rolls, the batch should return to the cooler, slower roll when the addition of compounding ingredients is complete.

The operation of extrusion is discussed in several chapters in this volume, so only a few general comments are made here. An extruder is a standard piece of equipment used in rubber and plastic processing for such operations as injection and blow molding, for processing thermosetting resins, and for the hybrid process of injection blow molding. The apparatus is fed room-temperature resin in the form of beads, pellets, or powders; or if rubbers are being processed, the feed material may be in the form of particulates or strips. The unit converts the feedstock into a molten polymer at sufficiently high pressure to enable the highly viscous melt to be forced through a nozzle into the mold cavity (injection type molding) or through a die (for example, blow molding or continuous extrusion of articles). In the initial portion of the extruder the polymer is conveyed along the extruder barrel and is compressed. The material is then heated until soft, eventually reaching a molten state. As fresh feed material enters, heat transfer takes place between the molten fluid and solid polymer. Once in the molten state, the extruder acts like a pump, transferring the molten polymer through the extruder channel, building up pressure prior to flow through the discharge nozzle or die. The principal components of a single-screw extruder are illustrated in Figure 4-16. The machine has a motor drive, a gear train, and a screw that is keyed into the gear-reducing train. The fluid layers between the screw flights and barrel wall maintain the screw balanced and centered.

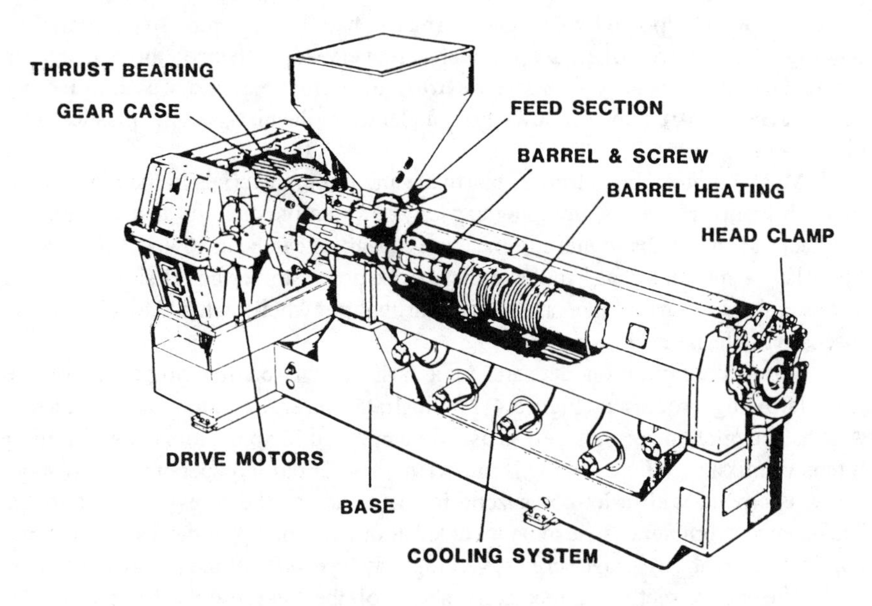

Figure 4-16 Features of a single-screw extruder

Units are equipped with continuous variable speeds, and barrels are often electrically heated, usually by band heaters with either on-off or proportional control. The barrel can be zoned according to the number of controllers on the heater bands. Depending on the application and type of service, the screw may be cored to allow for heating or cooling.

A die can be attached at the end of the extruder. Die designs range from very simple geometries such as an annulus for pipe and tubing profiles, to very complex faces such as the rubber seals used as glass run channels around the windows of automobiles.

Older extruders used in manufacturing various consumer articles are equipped with minimal instrumentation. Standard instrumentation usually consists of a pressure gauge at one point along the barrel (usually at the head) and a thermocouple in the hot-melt region. In noncritical operations, the operator will monitor pressure, temperature, and screw speed. Mass flow rates are typically monitored by the sample weight-time method.

In operations requiring close tolerance on extruded articles, a greater degree of instrumentation is employed. Usually, several pressure transducers and thermocouples along the barrel are used to ensure uniform extrusion and to control barrel and stock temperatures. Some designs may include thermocouples on the screw to monitor and control conveying flights.

There are numerous types of extruders, the most common of which is screw extrusion, which can be of the single or twin type. Another common type is the plunger or ram-type extruder. Materials such as TFE Teflon and ultrahigh-molecular-weight polyethylene are normally handled by ram extrusion. The melting temperatures of these polymers are very high; hence, these materials cannot be pumped readily as in screw extrusion. In this chapter discussions largely cover screw extrusion of low- to moderate-molecular-weight plastics and elastomers.

Many commercial extruders plasticate and pump materials in the range of 10 to 15 lb/hr-hp. However, pumping capacity is a relative quantity that depends on the material. In adiabatic mixing, machine capacities can be as low as 3 to 5 lb/hr-hp. Also, a machine that handles a thermoplastic elastomer could show as much as threefold- to fourfold increase in mass throughput when switched to a low-melt-viscosity material such as nylon.

Screw configuration depends to a large extent on the properties of the material being processed. Figure 4-17 illustrates several common arrangements. A constant-pitch metering screw is usually employed in applications not requiring intensive mixing. Where mixing is important, say for color dispersion, a two-stage screw equipped with a let-down zone in the center of the screw is appropriate. Turbulence promoters can also be included at or near the tip of the screw. In some applications with two-stage screws, venting may be needed at the let-down section.

The screw section immediately ahead of the gear train acts as a solids-metering or feed zone. It is characterized by a deep channel between the root of the screw and the barrel wall. The plasticating zone is ahead of the solids-metering zone. This is a transition region where the channel narrows. The purpose of this zone is to provide intense friction between solids and a region for the melting of the polymer to take place. Near the tip of the screw is the melt-metering zone, where pressure builds up. In this region the polymer melt is essentially homogenized and raised to the proper temperature for extrusion of the article.

The action of an extruder is analogous to that of a positive-displacement pump. The flight depth along the screw (that is, the ratio of the solids-metering channel depth to the melt-pump channel depth) is known as the compression ratio. The purpose of screw flights is to enable the screw to transport polymer down the barrel. The pitch angle of the flights again depends on the type of material handled. Many elastomer applications employ a general-purpose screw that is of a constant pitch (that is, flight equals the diameter). The pitch angle of this single-flighted screw is usually 17.61°. Typical extruder specifications are as follows:

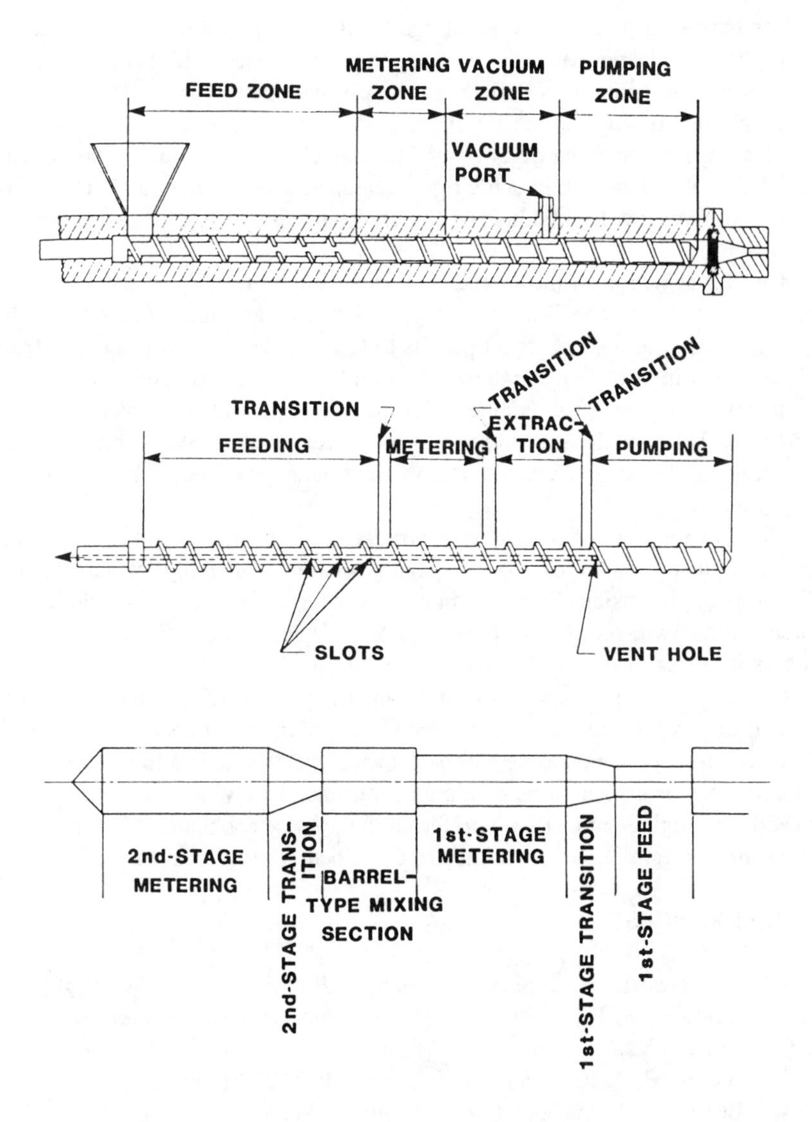

Figure 4-17 Screw profiles and cross section of vented barrel extruder

Compression ratio: 2:1 to 6:1 (for materials ranging from LDPE to some nylons)
Pitch angle: 12° to 20°
Length to diameter (L/D) 16:1 to 36:1 (low 40s typically for easily melting/flowing polymers requiring high mixing and venting)
Extrusion pressures: 10,000 to 300,000 psi

For extrusion pressures, low ratings (≤ 100,000 psi) are usually sufficient for many thermoplastic materials. The upper limit typifies FEP Teflon.

The lead on a screw is defined as the distance between the flights. As an approximation it is equal to the ID of the barrel for a single-flighted screw. The radial clearance between the flight tip and the barrel is tight (usually 0.001 in./in. of barrel ID). The reason for such a tight clearance is that if the gap is too great, the material may flow back along the barrel, resulting in a loss of melt-pumping zone capacity.

The initial region of the extruder plays an important role in the machine's overall operation. It consists of a hopper or feed arrangement and a solid feedstock conveying region. Its purpose is to transfer the cold polymer feed from the feed hopper into the barrel, where it is initially compressed. This compression forces air out between the interstices of resin pellets or rubber chunks (air being expelled back through the hopper) and breaks up lumps and polymer agglomerates. This action creates a more homogeneous feedstock that can be readily melted.

There is a variety of twin-screw extruder designs employed throughout the polymers industry, with each type having distinct operating principles and applications in processing. Designs can generally be categorized as corotating and counterrotating twin-screw extruders. Eise et al. (20) has classified twin-screw extruders in terms of the mechanisms of operation.

Molding operations are the next group of processing methods, generally divided into three techniques: compression, transfer, and injection. Each case requires a different compounding approach to achieve efficient mold fill. Although all EPDMs may be used in these molding operations, the low-Mooney, fast-curing (high and very high diene) EPs are preferred in the more sophisticated transfer and injection processes. These operations are described elsewhere in the volume.

4.6 REFERENCES

1. Andrews, E. H., and A. J. Walsh, *J. Polym. Sci.*, 33: 39 (1958).
2. Baldwin, F. P., and G. Ver Strate, *Rubber Chem. Technol.*, 45: 709-881 (1972).
3. Baranival, K. C., *Makromol. Chem.*, 100: 242 (1967).
4. Beatty, I. R., *Rubber Chem. Technol.*, 37: 1341 (1964).
5. Boonstra, B. B., *J. Appl. Polym. Sci.*, 11: 389-416 (1967).

6. Boonstra, B. B., in *Rubber Technology*, ed. M. Morton. New York: Litton, 1973.
7. Bridgwater, J., and A. M. Scott, in *Handbook of Fluids in Motion*, eds. N. Cheremisinoff and R. Gupta. Ann Arbor, MI: Ann Arbor Science Pub., 1983.
8. Burgess, K. A., S. M. Hirshfield, and C. A. Stokes, *Rubber Age*, 98(9): 85 (1965).
9. Burgess, K. A., S.S. Thune, and E. Palmese, *Rubber World*, 149(4): 34 (1974).
10. Carley, J. F., *Mod. Plast.*, 60(2): 77 (1971).
11. Cheremisinoff, N. P., *Polymer Mixing and Extrusion Technology*. New York: Marcel Dekker, 1987.
12. Cheremisinoff, P. N., *Carbon Adsorption for Pollution Control*, Englewood Cliffs, NJ: PTR Prentice Hall, 1993.
13. Chung, C. T., W. J. Hennessey, and M. H. Tusim, *Polym. Eng. Sci.*, 17: 9 (1977), 14.
14. Corish, P. J., *Rubber Chem. Technol.*, 44(3): 814 (1971).
15. Corish, P. J., and B.D.W. Powell, *Rubber Chem. Technol.*, 47(3): 481 (1974).
16. Cotten, G. R., paper 1, Rubber Division Meeting, American Chemical Society, Cleveland, Ohio (1975).
17. Cotten, G. R., paper presented at Rubber Division Meeting, American Chemical Society, Los Angeles (April 1985).
18. Cotten, G. R., *Rubber Chem. Technol.*, 5: 118 (1984).
19. Dekker, J., *Kunststoffe*, 66: 130 (1976).
20. Eise, K., H. Werner, H. Herrmann, and U. Burkhardt, in *Advances in Plastics Technology*, vol. 1, no. 2. New York: Van Nostrand, 1981.
21. Flory, R., *Principles of Polymer Chemistry*. Ithaca, NY: Cornell University Press, 1952.
22. Gessler, A. M., *Rubber Chem. Technol.*, 43(5): 943 (1970).
23. Gessler, A. M., W. M. Hess, and A. I. Medalia, *Reinforcement of Elastomers with Carbon Black, Plastics and Rubber Processing Paper 1164* (1978).
24. Hawkins, W. L., *Rubber Plast. Wkly. London*, 142: 291 (1962).
25. Hawkins, W. L., *Rubber Plast. Wkly. London*, 34: 1134 (1961).
26. Johnson, P. S., *Rubber Chem. Technol.*, 56: 3 (July-August 1983).
27. Kacir, L., and Z. Tadmor, *Polym. Eng. Sci.*, 12: 387 (1972).
28. Keller, R. C., paper presented at Rubber Division Meeting, American Chemical Society, Los Angeles (April 1985).
29. Klein, I., and D. I. Marshall, *SPE J.*, 21: 1376 (1965).
30. Kresge, E. N., and C. Cozewith, paper presented at Rubber Division Meeting, American Chemical Society, Cleveland, Ohio (1984).
31. Van Krevelen, D., *Properties of Polymers*. New York: Elsevier, 1972.
32. Lee, M.C.H., *J. Appl. Polym. Sci.*, 29: 499 (1984).

33. Leigh-Dugmore, C. H., *Rubber Chem. Technol.*, 29: 1303 (1956).
34. Marshall, D. I., I. Klein, and R. H. Uhl, *SPE J.*, 21: 1192 (1965).
35. Medalia, A. I., *Rubber Chem. Technol.*, 34: 1134 (1961).
36. Munstedt, H., *Kunststoffe*, 68: 92-98 (1978).
37. Nichols, R. J., *Kunststoffe German Plastics*, vol. 9. Munich: Carl Hanser Verlag, 1982.
38. Nielson, L. E., in *Mechanical Properties of Polymers and Composites*, vol. 2. New York: Marcel Dekker, 1974.
39. Payne, A. R., and W. F. Watson, *Rubber Chem. Technol.*, 36: 147 (1963).
40. Scholte, Th., in *Developments in Polymer Characterization*, vol . 4, ed. J. Dawkins. England, Applied Science Pub., 1983.
41. Scott, C. E., and F. J. Eckert, *Rubber Chem. Technol.*, 39(3): 533 (1966).
42. Sezna, J. A., *Elastomerics*, 16: 8 (August 1984).
43. Slade, P. E., *Polymer Molecular Weights, Part 1*. New York: Marcel Dekker, 1975.
44. Smallwood, H., *J. Appl. Phys.*, 15: 758 (1944).
45. Smith, W. R., and G. A. Kasten, *Rubber Chem. Technol.*, 44: 1287 (1970).
46. Squires, P. H., *Soc. Plast. Eng. Trans.*. 4(1): 7 (1964).
47. Tadmor, Z., D.I. Marshall, and I. Klein, *Polym. Eng. Sci.*, 6: 185 (1966).
48. Tadmor, Z., and E. Broyer, *Polym. Eng. Sci.*, 12: 378 (1972).
49. Tadmor, Z., and I. Klein, *Engineering Principles of Plasticating Extrusion*. New York: Van Nostrand (1970).
50. Thorne, J. L., *Plastics Process Engineering*. New York: Marcel Dekker, 1979.
51. Turetsky, S. B., P. R. Van Bushirk, and P. F. Gurberg, *Rubber Chem. Technol.*, 49: 1 (1976).
52. Ver Strate, G., in *Encyclopedia of Polymer Science and Engineering*. New York: McGraw-Hill, 1986.
53. Walters, M. H., and D. N. Keyte, *Trans. IRI*, 38: 40 (1962).
54. Zavadsky, E., *Chem. Vlakna*, 28: 160 (1978).
55. Zavadsky, E., J. Karnis, and V. Pechoc, *Rheol. Acta*, 21: 470-474 (1974).

5 DATA AND GENERAL INFORMATION ON POLYMER BLENDS

5.1 OVERVIEW

This chapter provides general information, properties, and a market breakdown of polymer blends. The specific advantages of one-phase behavior are:

1. Assured mechanical compatibility.
2. No problems during high-shear rate processing (for example, injection molding). Two-phase blends generally have poor weld line strength.
3. No problems with segregation of additives in a particular phase (very important in elastomer blends).
4. Offers excellent permanence when incorporated as a plasticizer.
5. Heat distortion enhancements are possible.

The following are short descriptions on major polymer-blend systems.

Nitrile Rubber - PVC: butadiene/acrylonitrile rubber (23 to 43 percent An) is miscible with PVC over the entire concentration range.
Initial reference: U.S. Patent No. 2,297,194 (1942).
In commercial use since early 1940s.

Uses:

Vulcanized nitrile rubber blends (uses plasticized PVC)

1. printing roll curves
2. gaskets
3. fuel hose covers
4. cable jackets
5. conveyor belts

Permanently plasticized PVC

1. jacketing and low-voltage insulation for wire and cable
2. nonmigratory plasticizer for fatty food contact
3. pond liners for oil containment

Commercial Miscible Blends/Poly(vinyl chloride): high molecular weight permanent plasticizers include:

1. nitrile rubber
2. E/VA copolymers (65% VA)
3. E/VA/CO elvaloy (Du Pont)
4. thermoplastic polyurethanes based on polyester soft blocks that is, poly(ϵ-caprolactone)
5. hytrel (Du Pont) poly (butylene terephthalate)-polytetrahydrofuran $(AB)_N$ block copolymers

Noryl/PPO-Polystyrene: is miscible with polystyrene over entire concentration range. The structure is

$$\left[-C_6H_2(CH_3)_2-O-\right]_n$$

Specific attributes:

price/performance compromise between constituents
PPO improves ability to flame retard polystyrene
PPO improves heat distortion temperature
polystyrene offers improved processability and lower cost
Refer to U.S. Patent No. 3,383,435 (1968) assigned to General Electric Co.
Noryl Miscible Blend of Polystyrene and

$$\text{PPO}\left[-C_6H_2(CH_3)_2-O-\right]_n$$

Typical applications:

1. Dishwasher parts: hydrolytic resistance and low creep.
2. Computer terminal housings: flammability rating.
3. Hair dryer components: heat distortion temperature and flammability rating.
4. Hospital furniture: flammability rating.
5. Humidifier parts: hydrolytic resistance and flammability rating.

6. Automotive dashboard: heat distortion temperature and impact strength.
7. Automotive grilles and trim: ability to be plated and heat distortion temperature.
8. Curling irons: heat distortion temperature and flammability rating.
9. Vacuum, cleaner housings: flammability rating.
10. Plumbing fixtures: heat distortion temperature and ability to be plated.

High Heat ABS - azeotropic compositions of α-methyl styrene/acrylonitrile (69/31 by weight) (T_g =128°C) and styrene/acrylonitrile (72/28 by weight) (T_g = 100°C) are miscible.

α-mS/An addition to ABS yields a higher heat distortion temperature than ABS.

References: U.S. Patent No. 3,010,936 (1961) assigned to Borg-Warner Corporation; Slocombe, R.J., *J. Poly. Sc.*, 26, 9 (1957).

Polymers miscible with poly(vinyl chloride) - The following is a list of polymers that are miscible with PVC:

	Polymer	Comments
1.	Butadiene/acrylonitrile copolymers	23-45% acrylonitrile
2.	Ethylene/vinyl acetate (E/VA) copolymers	65-70% VA
3.	Poly (ϵ-caprolactone)	
4.	E/VA/SO_2 terpolymers	Broad composition range
5.	E/VA/CO terpolymers	Commercially available as a permanent plasticizer
6.	E/EA (Ethyl acrylate)/CO terpolymers	
7.	Syndiotactic PMMA	Miscible up to 1:1 PMMA: PVS molar ratio
8.	α-Methyl styrene/acrylonitrile/copolymers	69/31 by weight
9.	α-Methyl styrene/acrylonitrile/methyl methacrylate terpolymer	60/20/20 by weight
10.	α-Methyl styrene/methacrylonitrile copolymer	57 mole % man
11.	Poly(butylene terephthalate)	
12.	Ethylene/N, N-dimethyl acrylamide copolymers	

Polymers miscible with cellulosics - The following is a list of polymers that are miscible with cellulosics.

Nitrocellulose - poly (ε-caprolactone)
Nitrocellulose - poly(methyl acrylate)
Nitrocellulose - Poly(methyl methacrylate)
Nitrocellulose - Poly(vinyl acetate)
Nitrocellulose - Ethyl cellulose
Nitrocellulose - Cellulose acetate proprionate
Cellulose acetate butyrate - E/VA (80% VA)
Cellulose butyrate, cellulose proprionate, cellulose acetate butyrate - poly (ε-caprolactone)

Polymers miscible with polystyrene or styrene copolymers - The following is a list of polymers that are miscible with polystyrene and styrene copolymers:

Atactic polystyrene - Isotactic polystyrene
Polystyrene - 2,6 dimethyl 1,4 phenylene oxide (PPO)
Polystyrene - Poly(vinyl methyl ether)
Polystyrene - Tetramethyl bisphenol a polycarbonate
Styrene/acrylonitrile san - Poly (ε-caprolactone) 72/28 copolymers
San (9 to 27 wt.% An) - PMMA
San (24% to An) - α-Methyl styrene/acrylonitrile (69/31) copolymer
San - Nitrocellulose

Polymers Miscible with Polyethers -

Poly(ethylene oxide) - poly (acrylic acid)
Poly(ethylene oxide) - poly (methacrylic acid)
Poly(ethylene oxide) - vinyl methyl ether/maleic anhydride copolymer
Poly(ethylene oxide) - carboxy methyl cellulose sodium salt
PEO Phenolic
PEO Phnoxy
PUME Phenolic

The following are typical mechanical properties of Poly(ethylene oxide)-Poly(acrylic acid) complexes.

	Poly (acrylic Acid)	**50/50 Blend**	**Poly (ethylene oxide)**
Tensile Modulus (psi)	400,000	200	50,000
T_g (°C)	100	5	-55
Water Extractables %	100	10	100

	Poly (acrylic Acid)	50/50 Blend	Poly (ethylene oxide)
Tensile Strength (psi)	9,000	1,500	2,100
Ultimate Elongation%	2	600	1,000

Polymers Miscible with Polyesters and Polycarbonates -

Poly (ϵ-Caprolactone)	- Phenoxy
Poly (ϵ-Caprolactone)	- Polyepichlorohydrin
Poly (ϵ-Caprolactone)	- Chlorinated polyether
Poly - (caprolactone)-	- Polycarbonate
Poly(butylene terephthalate)	- Phenoxy
Poly(ethylene terephthalate)	- Phenoxy
Poly(1,4 cyclohexylene dimethylene tere/iso phthalate)	- Phenoxy
Poly(ethylene terephthalate)	- Polycarbonate
Poly(1,4 cyclohexylene dimethylene tere/iso phthalate)	- Polycarbonate

Polymers Miscible with Polyamides -

Nylon 6	- Ethylene/acrylic acid (14 wt. % AA) copolymer
Nylon 6,6	- Polyamide from M-phenylene diamine and adipic acid
Nylon 6,6	- Polyamide from isophthalamide

Polymers Miscible with Poly(vinyl acetate) -

Poly(vinyl acetate)	- Poly(vinyl nitrate)
Poly(vinyl acetate)	- Methyl vinyl ether/maleic anhydride copolymer
Poly(vinyl acetate)	- Poly(vinylidene fluoride)

Polymers Miscible with Polyacrylate -

Syndiotactic PMMA	- Isotactic PMMA
Isotactic PMMA	- Syndiotactic poly(methacrylic acid)
PMMA	- Poly(vinylidene fluoride)
Poly(ethyl methacrylate)	- Poly(vinylidene fluoride)
Poly(methyl acrylate)	- Poly(vinyl acetate)
Poly(methyl acrylate)	- Poly(vinylidene fluoride)
Poly(ethyl acrylate)	- Poly(vinylidene fluoride)
Poly(isopropyl acrylate)	- Poly(isopropyl methacrylate)

Polymers Miscible with Unsaturated Hydrocarbon-Based Polymers -

Natural Rubber	- Polybutadiene (high 1,2 content)
CIS 1,4 Polybutadiene	- Styrene/butadiene copolymer
Chloroprene (neoprene)	- Styrene/butadiene copolymer
Chloroprene	- Nitrile rubber (18% An)
Nitrile Rubber (18% An)	- Nitrile rubber (40% An)
SBR (16% Styrene) Rubber	- SBR (23.5% styrene) rubber
SBR (37.5 Styrene) Rubber	- SBR (50% styrene) rubber

Miscellaneous Miscible Polymer Blends -

Poly(butene-1)	- Polypropylene
Chlorinated Rubber (65% CL)	- Ethylene/vinyl acetate copolymers (28-40 wt. % VA)
Poly(vinylidene fluoride)	- Poly(vinyl methyl ketone)
Chloroprene	- Phenolic (T-butyl phenol based)
Nitrile Rubber	- Phenolic (T-butyl phenol based)

Block Copolymers with Miscible Block Components -

Polystyrene	- Poly (α-methyl styrene) AB, ABA
Polysulfone	- Polycarbonate $(AB)_N$ 5,000/5,000 M 17,000/
Polysulfone	- Phenoxy $(AB)_N$
Polysulfone	- Poly(butylene terephthalate) $(AB)_N$ also low
Polycarbonate	- Poly(ethylene oxide) $(AB)_N$
Polycarbonate	- Poly (ϵ-Caprolactone) $(AB)_N$
Poly(Butylene)	- Polytetrahydrofuran $(AB)_N$

(terephthalate) similar to segmented polyurethanes

Block Copolymers (Miscible Blocks) -

Hytrel: Poly(butylene terephthalate)-poly(tetrahydrofuran) $(AB)_N$ block copolymer.
PBT and PTHF blocks are miscible in the amorphous phase.
PBT crystallize to yield a thermoplastic elastomer.
Thermoplastic polyurethanes:

In commercial thermoplastic polyurethanes partial miscibility occurs between the polyester (or polyether) blocks and the amorphous component of the urethane block:

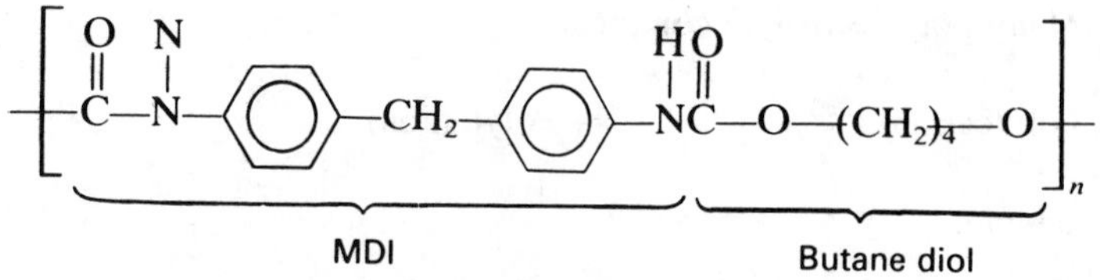

$$\left[\overset{O}{\overset{\|}{C}}-\overset{N}{\overset{|}{N}}-\langle\bigcirc\rangle-CH_2-\langle\bigcirc\rangle-\overset{H}{\overset{|}{N}}\overset{O}{\overset{\|}{C}}-O-(CH_2)_4-O\right]_n$$

MDI | Butane diol

Graft Copolymer with Miscible Graft Components

Nitrocellulose-$_G$-poly(methyl acrylate)
Nylon 6-$_G$-poly(ethylene oxide)
Nylon 6-$_G$-ethylene/acrylic acid copolymers
Poly(ethylene/propylene/1,4hexadiene)-G-polystyreneandpoly(isobutylene) (only elastomeric components miscible)
Poly(ethylene/propylene/1,4 hexadiene)-G-polystyrene and poly (α-methyl styrene) (only graft components miscible)

Miscibility in Interpenetrating Networks -

Polyurethane - Polyester
Polyurethane - Epoxy
Polyurethane - Polyacrylate
Poly(N-Butyl acrylate) - Poly(ethyl methacrylate)
Poly(Ethyl Acrylate) - Poly(methyl methacrylate)

Frisch, H. L., K. C. Frisch, and D. Klempner, *Poly. Eng. Sci.*, 14, 646 (1974).

Isomorphic Polymer Blends -

Poly(Vinyl Fluoride) - Poly(vinylidene fluoride)
Poly(Isopropyl Vinyl Ether) - Poly(sec-butyl vinyl ether)
Poly(4-Methyl Pentene) - Poly (4-methyl hexene)

Miscible Blends of Water-Soluble Polymers and Polyelectrolyte Complexes -

Poly(Vinyl Pyrrolidone) - Poly(acrylic acid)
Poly(Vinyl Pyrrolidone) - Poly(methacrylic acid)
Poly(Ethylene Imine) - Carboxymethyl Cellulose
Poly(Methacrylic Acid) - Poly-L-Lysine
Poly(Methacrylic Acid) - Poly(N,N-dimethlyamino ethyl methacrylate)
Poly(Ethylene Imine) - Poly(acrylic acid)
Poly(Ethylene Imine) - Poly (methacrylic acid)

Many polyelectrolyte complexes

Poly(cation) - Poly(anion)

For example:

Poly(sodium styrene sulfonate)
Poly(vinylbenzyl trimethyl ammonium chloride)

Polyelectrolyte Complexes

Poly(vinyl benzyl trimethyl ammonium chloride) - sodium polystyrene sulfonate. Ioplex resins and diaflo ultrafiltration membranes (Amicon Corporation).

Potential applications:

1. Hemodialyzer (artificial kidney) membranes.
2. Hemooxygenator (artificial lung) membranes.
3. Battery separators.
4. Fuel cell membranes.
5. Electrically conductive and antistatic coatings.
6. Photoresistant sheets.
7. Hydrogels as a blood "compatible" surface.
8. Microencapsulation systems.
9. Wound and burn dressings.
10. Soft contact lenses and cornea substitutes.
11. Matrix for slow release of implantable drugs.

5.2 MARKETS AND END USES

This section contains market breakdown information on the end uses and market growth patterns of polymer blends and alloys. Table 5-1 provides an industry characterization of specialty polymer blends and alloys. For perspective, Figure 5-1 provides a conceptual description of the differences between polymer alloys and blends.

Table 5-1 Specialty Polymer Blends and Alloys Industry Characterization

U.S. demand 1987	550 million pounds
Average price	$1.50 a pound
Forecast growth	8% a year

Table 5-1 Specialty Polymer Blends and Alloys Industry Characterization (continued)

Business focus	Marketing
Driving force	Suppliers
Life-cycle stage	Growth
Technical support	High
Product differentiation	High
End-use know-how	High
Capital intensity	Low
Cyclicality	Low
Globalization	Moderate
Key skills	Product development Applications development Marketing ability

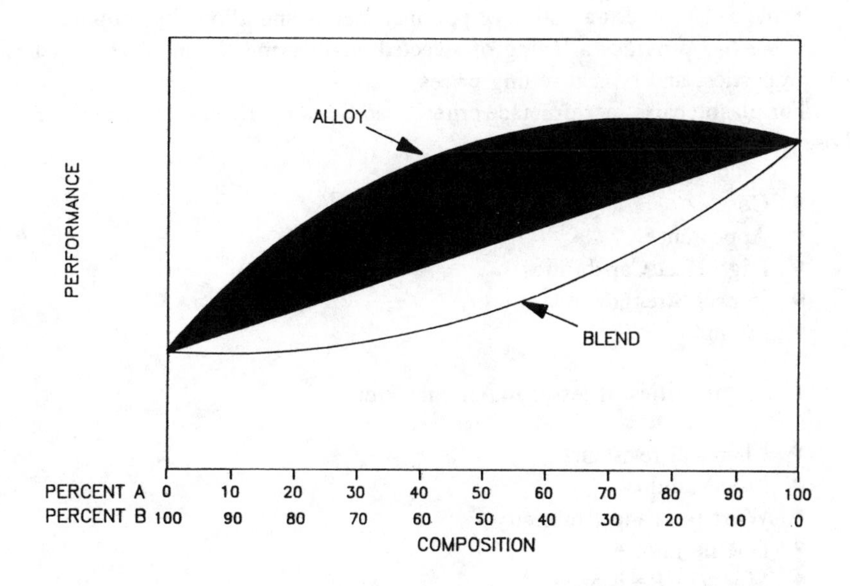

Figure 5-1 Illustrates the definition of polymer blends and alloys

Table 5-2 provides a listing of specialty polymer blends and alloys, along with their trade names.

Note that we can classify commercial polymer blends by means of their property improvements:

Additive blends	**Property improvement blends (impact)**	**Synergistic blends**
ABS/PC	Modified acetal	ABS/nylon
ABS/PVC	Modified inomer	
ABS/SMA	Modified nylon	
PBT/PET	Modified PBT	
PC/PBT	Modified PET	
PC/PET	PC/PE	
PC/SMA	PP/EPDM	
PPO/PS		
PPO/nylon		
PC/nylon		
Nylon/PP		
PP/nylon		

A breakdown of the consumption of major polymer blends and alloys is given in Figure 5-2. Further information is given in Table 5-3.

Table 5-4 provides a listing of polymer blends and alloys by supplier.

Table 5-5 provides a listing of selected engineering thermoplastic grades, their properties, and typical selling prices.

For plastic parts, performance criteria can be categorized into several areas. Those of major importance are:

- Cost
- Appearance
- High Heat Capabilities
- Impact Strength
- Strength

Those properties of lesser importance are:

- Chemical resistance
- Processability
- Wear resistance, lubricity
- Ductile failure
- Moisture Resistance
- Flame Resistance

Conceptually the properties of PPO/PS alloys can be summarized as in Figure 5-3. The suppliers of PPO/PS blends by application in millions of pounds are as follows:

Table 5-2 Speciality Polymer Blends and Alloys with Trade Names

Product	Supplier	Trade Name	Grades
ABS/nylon	Borg-Warner	ELEMID	RMI
ABS/nylon	Monsanto	TRIAX	All
ABS/PC	Borg-Warner	PROLOY	All
ABS/PC	Borg-Warner	CYCOLOY	All
ABS/PC	Dow	PULSE	All
ABS/PC	Mobay	BAYBLEND	All
ABS/PVC	Borg-Warner	CYCOVIN	All
ABS/PVC	Amoco	MINDEL	A-670, A-650
ABS/PVC	Monsanto	LUSTRAN	865, 860
ABS/PVC	Schulman	POLYMAN	507, 509, 511
ABS/PVC	Shuman	–	780
ABS/SMA	Monsanto	CADON	All
Modified acetal	BASF	ULTRAFORM	N2640X
Modified acetal	Du Pont	DELRIN	ST-100,T-500
Modified acetal	Hoechst Celanese	DURALOY	1000, 1100, 1200, 21
Modified acetal	Hoechst Celanese	CELCON	C-400, C-401
Modified ASA	BASF	TERBLEND	SKR2861
Modified ASA	General Electric	GELOY	13320, 1221, 1220
Modified ionmer	Du Pont	SURLYN	HP series
Modified nylon	Allied Chemical	CAPRON	8250, 8252, 8350
Modified nylon	BASF	ULTRAMID	KR4430, B3L, KR4645
Modified nylon	Du Pont	ZYTEL	ST series, 408
Modified nylon	Emser Industries	GRILON	A28 Series
Modified nylon	Hoechst Celanese	CELANESE	N-297, N-303
Modified nylon	INP	THERMOCOMP VF	All
Modified nylon	Nylon Corp. of America	NYCOA	1417, 2001
Modified nylon	Thermofil	–	All
Modified nylon	Wilson Fiberfil	NYAFIL	TN
Modified PBT	BASF	ULTRADUR	KR4070, KR4071
Modified PBT	Comalloy	VOLEX	430, 431
Modified PBT	Comalloy	HILOY	431, 432, 433, 434
Modified PBT	Comalloy	COMTUF	431
Modified PBT	General Electric	VALOX	340
Modified PBT	Hoechst Celanese	VANDAR	4424F
Modified PBT	Mobay	POCAN	S1506

Table 5-2 Speciality Polymer Blends and Alloys with Trade Names (*Continued*)

Product	Supplier	Trade Name	Grades
Modified PET	Comalloy	VOLEX	460, 461, 462
Modified PET	Comalloy	HILOY	441, 461, 462, 463
Modified PET	Comalloy	COMUIF	461, 462, 464, 464
Modified PET	Du Pont	RYNITE SST	All
Nylon/PP	Dexter	DEXLON	All
PBT/PC	Comalloy	–	–
PBT/PET	Comalloy	HILOY	432
PBT/PET	General Electric	VALOX	860, 855, 815, 830
PBT/PET	Hoechst Celanese	CELANEX	5330
PC/nylon	Dexter	DEXCRAB	All
PC/PE	General Electric	LEXAN	191
PC/PE	Mobay	MAKROLON	T-7855-T-7955
PC/polyester	General Electric	VALOX	508m 553
PC/polyester	General Electric	XENOY	All
PC/polyester	Mobay	MARKOBLEND	UT
PC/polyester	Thermofil	R2	All
PC/SMA	Arco Chemical	ARLOY	All
PC/TPU	Mobay	TEXIN	902, 3203
PET/PC	Comalloy	–	–
PET/PSF	Amoco	MINDEX	B-390, B-322
PPE/PS	General Electric	NORYL	All
PPO/nylon	BASF	ULTRANYL	All
PPO/nylon	General Electric	NORYL GTX	All
PPO/nylon	BASF	ULTRANYL	All
PPO/PBT	General Electric	NORYL GTX	All
PPO/PS	BASF	LURANYL	All
PPO/PS	Borg-Warner	PREVEX	All
PP/EPDM	Dexter	ONTEX	All
PP/EPDM	Ferro	FERROFLEX	All
PP/EPDM	Schulman	POLYTROPE	All
PP/EPDM	Teknor Apex	TELCAR	All
PP/nylon	Dexter	DEXPRO	All
PVC/acrylic	Rohm & Haas	KYDEX	All
PVC/acrylic	Polycast Technology	–	All
PVC/urethane	Dexter	VYTHENE	All
SAN/EPDM	Dow	ROVEL	701, 705, 401, 501
SMA/HIPS	Arco Chemical	DYLARK	238

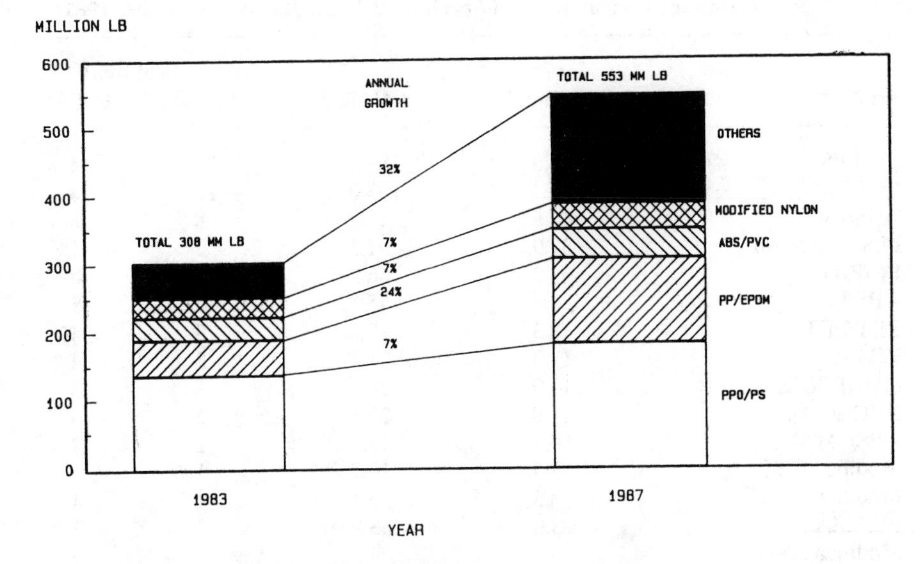

Figure 5-2 Consumption of major polymer blends

End Use	General Electric	Borg-Warner	Total
Business machines	70.2	12.5	82.7
Appliances	35.8	0.4	36.2
Automotive	31.0	3.4	34.4
Electronics	20.0	2.0	22.0
Medical	5.2	0.4	5.6
Transportation	5.4	—	5.4
Total	167.6	18.7	186.3

Another very lucrative market is PP/EPDM blend applications. The sales of P/EPDM by supplier based on 1987 figures are given in Table 5-6. The predominant market is automotive. Table 5-7 provides a listing of automotive PP/EPDM applications by process.

Table 5-8 provides estimated consumptions of PC/PBT blends.

Table 5-9 provides estimated consumptions of modified nylon.

Table 5-10 lists major suppliers of modified nylon.

Table 5-11 provides the estimated consumption of ABS/PVC alloys.

Table 5-3 Estimated Consumption of Specialty Polymer Blends and Alloys, 1987

Product	Million lb	$ Million	% of total Pounds	% of total Dollars
PPO/PS	186.3	283.1	34	34
PP/EPDM	125.1	133.0	23	16
ABS/PVC	42.6	49.0	8	6
Modified nylon	37.9	91.3	7	11
PC/PBT	34.3	66.9	6	8
ABS/PC	30.5	45.1	6	5
PBT/PET	18.3	29.6	3	4
PC/PE	17.3	31.7	3	4
SAN/EPDM	15.0	24.8	3	3
PVC/acrylic	11.9	23.8	2	3
ABS/SMA	10.7	15.0	2	2
Modified PBT	5.9	10.7	1	1
Modified acetal	3.5	6.4	1	1
PC/SMA	3.0	3.9	1	–
Modified ASA	3.0	4.6	1	1
PC/PET	2.0	3.9	–	–
Modified ionomer	2.0	1.9	–	–
Modified PET	1.2	2.4	–	–
ABS/nylon	0.9	1.6	–	–
PSF/PET	0.6	1.8	–	–
ABS/PSF	0.6	1.6	–	–
PPE/nylon	0.6	1.1	–	–
PPO/PBT	[a]	[a]	–	–
Total	553.2	833.2	100	100

[a] Minor.

Table 5-4 Specialty Polymer Blends and Alloys by Supplier

Supplier	Product	Trade Name	Grades
Allied Chemical	Modified nylon	CAPRON	8250, 8252, 8350
Amoco	ABS/PSF	MINDEL	A-670, A-650
Amoco	PET/PSF	MINDEL	B-390, B-322
Arco Chemical	PC/SMA	ARLOY	All
BASF	Modified acetal	ULTRAFORM	N2640X
BASF	Modified ASA	TERBLEND	SKR2861
BASF	Modified nylon	ULTRAMID	KR4430, B3L, KR4645
BASF	Modified PBT	ULTRADUR	KR4070, KR4071
BASF	PPO/nylon	ULTRANYL	All
BASF	PPO/PS	LURNYL	All
Borg-Warner	ABS/nylon	ELEMID	RM1
Borg-Warner	ABS/PC	CYCOLOY	All
Borg-Warner	ABS/PC	PROLOY	All
Borg-Warner	ABS/PVC	CYCOVIN	K-20, K-25, K-29
Borg-Warner	PPO/PS	PREVEX	All
Comalloy	Modified PBT	COMTUF	431
Comalloy	Modified PBT	VOLEX	430, 431
Comalloy	Modified PBT	HILOY	431, 432, 433, 434
Comalloy	Modified PBT	VOLEX	460, 461, 462
Comalloy	Modified PBT	COMTUF	461, 462, 464, 464
Comalloy	Modified PBT	HILOY	441, 461, 462, 463
Comalloy	PBT/PC	–	–
Comalloy	PBT/PET	HILOY	432
Comalloy	PET/PC	–	–
Dexter	Nylon/PP	DEXLON	All
Dexter	PC/nylon	DEXCARB	All
Dexter	PP/EPDM	ONTEX	All
Dexter	PP/nylon	DEXPRO	All
Dexter	PVC/urethane	VYTHENE	All
Dow	ABS/PC	PULSE	All
Dow	SAN/EPDM	ROVEL	701, 705, 401, 501
Du Pont	Modified acetal	DELRIN	ST-100, T-500
Du Pont	Modified ionomer	SURLYN	HP series
Du Pont	Modified nylon	ZYTEL	ST series, 408
Du Pont	Modified PET	RYNITE SST	All
Emser Industires	Modified nylon	GRILON	A28 series
Ferro Chemical	PP/EPDM	FERROFLEX	All

Table 5-4 Specialty Polymer Blends and Alloys by Supplier (*Continued*)

Supplier	Product	Trade Name	Grades
General Electric	Modified ASA	GELOY	1320, 1221, 1220
General Electric	Modified PBT	VALOX	340
General Electric	PBT/PET	VALOX	860, 855, 815, 830
General Electric	PC/PE	LEXAN	191
General Electric	PC/polyester	VALOX	508, 553
General Electric	PC/polyester	XENOY	All
General Electric	PPE/PS	NORYL	All
General Electric	PPO/nylon	NORYL GTX	All
General Electric	PPO/PBT	GEMAX	All
Hoechst Celanese	Modified acetal	DURALOY	1000, 1100, 1200, 21
Hoechst Celanese	Modified acetal	CELCON	C-400, C-401
Hoechst Celanese	Modified nylon	CELANESE	N-297, N-303
Hoechst Celanese	Modified PBT	VANDAR	All
Hoechst Celanese	PPT/PET	CELANEX	4424F
LNP	Modified nylon	THERMOCOMP VF	All
Mobay	ABS/PC	BAYBLEND	All
Mobay	Modified PBT	POCAN	S1506
Mobay	PC/PE	MAKROLON	T-7855, T-7955
Mobay	PC/polyester	MAKROBLEND	UT
Monsanto	ABS/nylon	TRIAX	All
Monsanto	ABS/PVC	LUSTRAN	865, 860
Monsanto	ABS/SMA	CADON	All
Nylon Corp. of America	Modified nylon	NYCOA	1417, 2001
Polycast Technology	PVC/acrylic	–	All
Rohm & Haas	PVC/acrylic	KYDEX	All
Schulman	ABS/PVC	POLYMAN	507, 509, 511
Schulman	PP/EPDM	POLYTROPE	All
Schuman	ABS/PVC	–	780
Teknor Apex	PP/EPDM	TELCAR	All
Thermofil	Modified nylon	–	All
Thermofil	PC/polyester	R2	All
Wilson Fiberfil	Modified nylon	NYAFIL	TN

Table 5-5 Selected Engineering Thermoplastic Grades and Properites

Resin	Denisty (g/cc)	Cost (c/cu in)	Flexural Modulus (psi *1,000)	Tensile strength (psi)	Izod impact (ft-lb/in)	HDT (F @264 psi)	Price ($/lb)
CYCOLAC L	1.02	4.3	270	5,000	7.5	188	1.18
DELRIN 100ST	1.34	9.1	200	6,500	17.0	194	1.89
DELRIN 500	1.42	8.4	410	10,000	1.4	255	1.64
DEXCARB 507 (PC/NY)	1.16	8.8	310	7,300	12.0	210	2.10[a]
DEXLON 632 (NY/PP)	1.35	8.8	1,000	24,000	1.9	480	1.80[a]
DEXPRO 136 (PP/NY)	1.14	4.5	720	13,500	1.8	330	1.10[a]
EKTAR (30% GR PP)	1.12	3.9	780	10,000	2.0	288	0.97
ELEMID RM1	1.09	7.0	300	6,910	18.7	182	1.77
GEMAX (PPO/PBT)	1.15	7.5	270	6,100	12.0	200	1.80
LEXAN 181	1.20	9.1	340	9,000	18.0	275	2.09
NORYL GTX 910	1.10	7.9	310	8,000	4.0	290	2.00
NORYL N190	1.08	6.4	325	7,000	7.0	190	1.63
PULSE 830	1.14	5.8	340	9,000	12.0	230	1.42
RYNITE 530	1.56	8.9	1,300	23,000	1.9	435	1.58
RYNITE SST35	1.51	10.6	1,000	15,000	4.4	428	1.95
VALOX 365 (impact)	1.31	9.9	325	6,000	12.0	250	2.10
VALOX 420 (30% FG)	1.53	9.9	1,100	17,300	1.6	410	1.80
VALOX 830 (PBT/PET)	1.54	9.0	1,000	15,000	1.5	380	1.61
XENOY 5220	1.21	8.7	296	7,700	13.3	210	1.99
ZYTEL 101	1.14	8.4	410	12,000	1.0	194	2.05
ZYTEL 70G33L	1.38	10.2	1,300	27,000	2.0	480	2.05
ZYTEL 71G33L	1.35	10.0	1,000	22,000	2.4	475	2.06
ZYTEL ST801	1.08	9.4	245	7,500	17.0	160	2.41
Minimum	1.02	3.9	200	5,000	1.0	160	0.97
Maximum	1.56	10.6	1,300	27,000	18.7	480	2.41

[a] Estimated

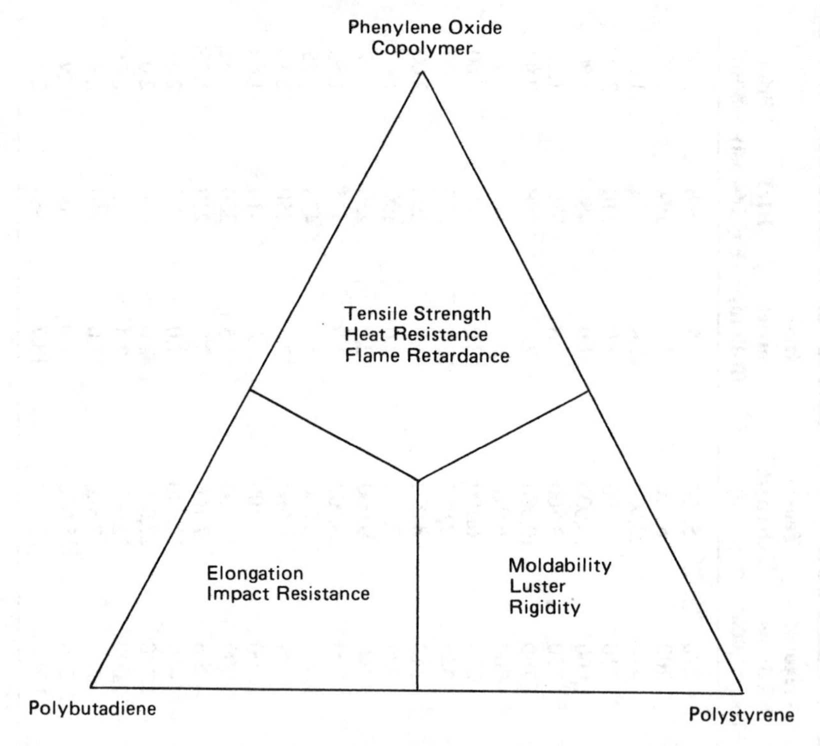

Figure 5-3 Illustrates properties of PPO/PS alloys

Table 5-6 Estimated Sales of PP/EPDM by Supplier, 1987

Supplier	Million lb	$ Million	% of total Pounds	% of total Dollars
RPI/Dexter	37.5	38.3	30	30
Himont	31.4	32.0	25	24
A. Schulman	15.0	16.0	12	12
Monsanto	13.5	15.1	11	11
Teknor Apex	5.8	5.9	5	4
Ferro	4.0	4.2	3	3
Thermofil	3.0	3.1	2	2
Other	14.9	18.4	12	14
Total	125.1	133.0	100	100

Table 5-7 Automotive PP/EPDM Applications by Process

Application	Process		
	Blow molding	Extrusion	Injection molding
Air dams	–	X	–
Air ducts	X	–	–
"A" pillars	–	–	X
Body side moldings	–	X	–
Bumper covers	–	–	X
Bumper end caps	–	–	X
Diesel barrier pad	–	–	X
Fender flares	–	–	X
Filler panels	–	–	X
Grilles	–	–	X
Primary wiring	–	X	–
Rocker panel	–	X	–
Rub strips	–	X	–
Scuff plates	–	–	X
Strut dust covers	–	–	X
Valence panels	–	–	X

Table 5-8 Estimated Consumption of PC/PBT Blends, 1987

End Use	Million lb	$ Million	% of total	
			Pounds	Dollars
Automotive	30.0	58.5	87	87
Business machines	2.6	5.1	8	8
Electronics	0.5	1.0	1	1
Power tools	0.2	0.4	a	a
Recreation	0.1	0.2	a	a
Other	0.9	1.7	4	4
Total	34.3	66.9	100	100

Table 5-9 Estimated Consumption of Modified Nylon, 1987

End Use	Million lb	$ Million	% of total	
			Pounds	Dollars
Automotive	26.4	63.6	69	69
Appliances	4.4	10.6	12	12
Recreation	3.6	8.7	10	10
Power tools	2.0	4.8	5	5
Electronics	1.5	3.6	4	4
Business machines	[a]	[a]	[a]	[a]
Transportation	[a]	[a]	[a]	[a]
Total	37.9	91.3	100	100

[a] Minor

Table 5-10 Major Suppliers of Modified Nylon, 1987

Supplier	Million lb	$ Million	% of total lb
Du Pont	27.9	67.4	74
Hoechst Celanese	2.7	6.5	7
Allied	1.8	4.3	5
Wilson-Fiberfil	1.8	4.3	5
Nycoa	1.3	3.1	3
Emser	0.8	1.9	2
LNP	0.5	1.2	1
Thermofil	0.3	0.7	1
Other[a]	0.8	1.9	2
Total	37.9	91.3	100

[a] Includes Wellman, Texapol, RTP, Ferro and others.

Table 5-11 Estimated Consumption of ABS/PVC Alloys, 1987

End Use	Million lb	$ Million	% of total	
			Pounds	Dollars
Automotive	28.5	32.7	67	67
Electronics	6.5	7.5	15	15
Appliances	1.5	1.7	3	3
Transportation	1.2	1.4	3	3
Business machines	0.8	0.9	2	2
Other	4.1	4.8	10	10
Total	42.6	49.0	100	100

Table 5-12 Consumption of ABS/PVC Alloys by Process, 1987

End Use	Extrusion	Injection	Calendering	Total
Automotive	–	5.1	23.4	28.5
Electronics	–	6.5	–	6.5
Appliances	–	1.5	–	1.5
Transportation	–	0.5	0.7	1.2
Business machines	–	–	0.8	0.8
Other	2.6	1.5	–	4.1
Total	2.6	15.1	24.9	42.6

Table 5-13 Estimated Sales of ABS/PVC Alloys by Supplier and Process, 1987

Supplier	Extrusion	Million lb Injection	Calendering	Total	% of total
Merchant					
Borg-Warner	0.6	10.1	–	10.7	25%
A. Schulman	–	4.0	–	4.0	9
Monsanto	2.0	0.8	–	2.8	7
Others	–	0.2	–	0.2	1
Total	2.6	15.1	–	17.7	42
Captive					
O'Sullivan	–	–	10.4	10.4	24
General Motors	–	–	8.0	8.0	19
Uniroyal	–	–	6.5	6.5	15
Total	–	–	24.9	24.9	58
TOTAL	2.6	15.1	24.9	42.6	100

Table 5-14 Estimated Consumption of ABS/PVC Alloys, 1987

End Use	Million lb	$ Million	% of total Pounds	Dollars
Business machines	12.9	20.5	42	46
Automotive	11.3	15.3	37	34
Electronics	2.6	4.2	9	9
Power tools	1.9	2.4	6	5
Appliances	1.5	2.3	5	5
Medical	0.3	0.4	1	1
Total	30.5	45.1	100	100

Table 5-15 gives estimated consumption of PBT/PET blends.

Table 5-16 gives the consumption of specialty polymer blends and alloys in the appliance industry. Applications for PPO/PS in the appliance industry include:

- Bearing housings
- Transformer housings

- Electrical outlet extenders
- Electrical boxes
- Vacuum cleaner parts
- Convection oven fan housings
- Transmissions (dishwashers and automatic washing machines)
- Air conditioner sump covers
- Motor housings (vacuum cleaners, dishwashers, automatic washing machines)
- Microwave oven parts
- Automatic washing machine motor pumps
- Dishwasher escutcheons

Table 5-17 provides estimated consumptions of specialty polymer blends and alloys in the appliance industry by application.

Table 5-18 provides estimated consumption of specialty polymer blends and alloys in the automotive industry. Table 5-19 provides a summary of applications and criteria for blends and alloys in automotive uses. Table 5-20 lists major suppliers of blends and alloys for this market sector. Table 5-21 summarizes consumption in the automotive sector by area of application.

Table 5-22 gives estimated consumption in the business machine industry.

Table 5-23 provides estimated consumption figures in the electronics industry.

Table 5-24 provides estimate consumption figures in the power tools industry.

Table 5-25 provides estimated consumption in the transportation industry. Table 5-26 lists consumption in the transportation industry by application.

Table 5-27 lists consumption in the recreation industry.

Table 5-28 lists consumption in the medical industry.

Table 5-29 lists sales of specialty polymer blends and alloys by major suppliers.

Table 5-15 Estimated Consumption of PBT/PET Blends, 1987

End Use	Million lb 1987	Million lb 1992	$ Million 1987	$ Million 1992	% Average annual growth, lb
Automotive	14.1	15.6	22.7	25.1	2
Appliances	2.9	3.2	4.7	5.3	2
Electronics	1.0	1.6	1.6	2.6	10
Power tools	0.3	0.3	0.5	0.5	—
Business machines	[a]	[a]	0.1	0.1	10
Total	18.3	20.7	29.6	33.6	2

[a] Minor

Table 5-16 Estimated Consumption of Specialty Polymer Blends and Alloys in the Appliance Industry, 1987

Blend/alloy	Million lb	$ Million	% of total Pounds	% of total Dollars
PPO/PS	36.2	55.0	69	66
Modified nylon	4.4	10.6	8	13
ABS/SMA	4.3	6.4	8	8
PBT/PET	2.9	4.7	6	6
ABS/PC	1.5	2.3	3	3
ABS/PVC	1.5	1.7	3	2
PP/EPDM	1.2	1.2	2	1
Modified acetal	0.6	1.1	1	1
Total	52.6	83.0	100	100

Table 5-17 Estimated Consumption of Specialty Polymer Blends and Alloys in the Appliance Industry by Application, 1987

	Million lb								
Blend/alloy	Vacuum cleaners	Refrig-erators	Washing machines	Dish-washers	Micro-waves	Ranges	Other	Total	% Of total
PPO/PS	8.1	3.8	1.5	2.0	2.0	0.9	17.9	36.2	72
Modified nylon	2.6	—	0.3	—	0.4	—	1.1	4.4	9
ABS/SMA	2.1	—	—	—	—	—	2.2	4.3	9
PBT/PET	—	—	—	—	—	[a]	2.9	2.9	2
ABS/PC	[a]	—	—	—	—	0.2	1.3	1.5	3
ABS/PVC	[a]	—	0.6	—	—	0.9	—	1.5	3
PP/EPDM	—	—	0.5	0.4	—	—	0.3	1.2	1
Modified acetal	0.2	—	—	0.4	—	—	—	0.6	1
Total	13.0	3.8	2.9	2.8	2.4	2.0	25.7	52.6	100

[a] Minor

Table 5-18 Estimated Consumption of Specialty Polymer Blends and Alloys in the Automotive Industry, 1987

Product	Million lb	$ Million	% of total	
			Pounds	Dollars
PP/EPDM	106.5	108.7	36	25
PPO/PS	34.4	52.3	12	12
PC/PBT	30.0	58.5	10	14
ABS/PVC	28.5	32.7	10	8
Modifed nylon	26.4	63.6	9	15
PC/PE	16.5	29.7	6	7
PBT/PET	14.1	22.7	5	5
ABS/PC	11.3	15.3	4	4
ABS/SMA	6.0	8.1	2	2
Modified PBT	5.2	9.4	2	2
SAN/EPDM	3.5	5.8	1	1
PC/SMA	3.0	3.9	1	1
Modified acetal	2.7	4.9	1	1
Modified ionomer	2.0	1.9	1	[a]
PC/PET	1.3	2.5	[a]	1
Modified PET	1.0	2.0	[a]	1
ABS/nylon	0.9	1.6	[a]	[a]
PPO/Nylon	0.5	0.9	[a]	[a]
Modified ASA	0.5	0.8	[a]	[a]
ABS/PSF	0.4	1.1	[a]	[a]
PSF/PET	0.3	0.9	[a]	[a]
PPO/PBT	[a]	[a]	[a]	[a]
Total	295.0	427.3	100	100

[a] Less than 1%.

Table 5-19 Application and Selection Criteria for Blends and Alloys in Automotive Applications, 1987

Area/Application	Impact strength Low temperature	Impact strength Normal	Flexural modulus	Heat distortion	Solvent resistance	Appearance	Electrial properties
Exterior							
Cowl Vents	M	M	M	H	H	H	L
Fascias/covers	H	H	H	H	H	H	L
Fenders	H	H	H	H	H	H	L
Grilles	H	H	H	H	H	H	L
Handles	M	M	H	M	M	M	L
Headlamp bezels	M	M	H	H	M	L	M
Mirror housings	H	H	M	H	H	H	L
Moldings	H	H	M	H	H	H	L
Wheel covers	H	H	H	H	H	H	L
Other	M	M	M	M	H	M	L
Interior							
Door mechanism	L	L	M	L	L	L	L
Door/window handle	L	L	H	M	L	M	L
Instrument panel	M	H	H	H	L	M	L
IP/door panel cover	L	L	L	H	L	H	L
Package shelf	L	L	M	M	L	H	L
Scuff plate trim	M	H	M	M	L	H	L
Steering column cover	L	M	H	M	L	H	L
Stoplight housing	L	L	H	H	L	H	M

L = low
M = medium
H = high

Table 5-19 Application and Selection Criteria for Blends and Alloys in Automotive Applications, 1987 (*Continued*)

Area/Application	Impact strength Low temperature	Impact strength Normal	Flexural modulus	Heat distortion	Solvent resistance	Appearance	Electrial properties
Underhood							
Air ducts	L	L	M	M	M	L	L
Engine control	L	L	M	H	H	L	H
Radiator fans	L	M	H	M	M	L	L
Radiator end tanks	L	L	H	H	H	L	L
Water pump housing	L	L	M	M	M	L	L
Other	L	L	M	M	M	L	L
Eletrical							
Ignition	L	L	M	H	M	L	H
Distributor cap	L	L	H	H	M	L	H
Switches	L	L	H	H	M	L	H
Connectors	L	L	H	H	N	L	H
Fuel System							
Emission cannister	L	L	M	M	H	L	L
Gas cap	L	L	H	M	H	L	L
Miscellaneous							
Front-end retainer	H	H	H	H	M	M	L
Lamp housing	M	M	H	H	M	M	M
Winshield wiper	M	M	H	M	M	L	L

Table 5-20 Major Suppliers of Polymer Blends and Alloys in the Automotive Industry, 1987

	Million lb									
Supplier	PP/EPDM	PPO/PS	ABS/PVC	PC/PBT	Modified nylon	PC/PE	PBT/PET	ABS/PC	Other	Total	% of total
General Electric	—	31.0	—	30.0	—	11.8	10.1	—	4.7	87.6	30
Dexter	37.5	—	—	—	—	—	—	—	—	37.5	13
Himont	30.3	—	—	—	—	—	—	—	—	30.3	10
Du Pont	—	—	—	—	20.1	—	—	—	3.9	24.0	8
Monsanto	8.2	—	0.5	—	—	—	—	—	6.5	15.2	5
A. Schulman	13.0	—	1.1	—	—	—	—	—	0.3	14.4	5
Borg-Warner	—	3.4	3.4	—	—	—	—	5.0	0.4	12.2	4
Hoechst Celanese	—	—	—	—	2.7	—	4.0	—	3.0	9.7	3
Mobay	—	—	—	—	—	3.5	—	3.5	1.3	8.3	3
Dow	—	—	—	—	—	—	—	2.8	3.5	6.3	2
Teknor Apex	5.8	—	—	—	—	—	—	—	—	5.8	2
Ferro	3.6	—	—	—	—	—	—	—	—	3.6	1
Arco	—	—	—	—	—	—	—	—	3.0	3.0	1
Other	8.1	—	23.5	—	3.6	1.2	—	—	0.7	37.1	13
Total	106.5	34.4	28.5	30.0	26.4	16.5	14.1	11.3	27.3	295.0	100

Table 5-21 Consumption of Specialty Polymer Blends and Alloys in the Automotive Industy by Area of Application, 1987

Million lb

Product	Exterior	Interior	Underhood	Other	Total
PP/EPDM	91.8	7.4	7.3	–	106.5
PPO/PS	9.2	23.3	1.1	0.8	34.4
PC/PBT	26.7	–	–	3.3	30.0
ABS/PVC	–	28.5	–	–	28.5
Modified nylon	12.9	1.5	4.6	7.4	26.4
PC/PE	–	16.5	–	–	16.5
PBT/PET	9.8	–	–	4.3	14.1
ABS/PC	–	8.2	–	3.1	11.3
ABS/SMA	0.6	5.4	–	–	6.0
Modified PBT	2.0	0.9	–	2.3	5.2
SAN/EPDM	0.5	3.0	–	–	3.5
PC/SMA	–	3.0	–	–	3.0
Modified acetal	0.4	1.3	–	1.0	2.7
Modified ionomer	2.0	–	–	–	2.0
PC/PET	–	–	–	1.3	1.3
Modified PET	–	–	1.0	–	1.0
ABS/Nylon	0.9	–	–	–	0.9
Modified ASA	–	–	–	0.5	0.5
PPO/Nylon	0.5	–	–	–	0.5
Other[a]	–	0.4	–	0.3	0.7
Total	157.3	99.4	14.0	24.3	295.0

[a] Includes ABS/PSF, PPO/PBT, PSF/PET, and others

Table 5-22 Estimated Consumption of Specialty Polymer Blends and Alloys in the Business Machine Industry, 1987

Blend/alloy	Million lb	$ Million	% of total Pounds	% of total Dollars
PPO/PS	82.7	125.7	83	83
ABS/PC	12.9	20.5	13	13
PC/PBT	2.6	5.1	3	3
ABS/PVC	0.8	0.9	1	1
Modified acetal	0.1	0.2	–	–
PBT/PET	[a]	0.1	–	–
Modified nylon	[a]	[a]	–	–
PP/EPDM	[a]	[a]	–	–
Total	99.1	152.5	100	100

[a] Minor

Table 5-23 Estimated Consumption of Specialty Polymer Blends and Alloys in the Electronic Industry, 1987

Product	Million lb	$ Million	% of total Pounds	Dollars
PPO/PS	22.0	33.4	61	61
ABS/PVC	6.5	7.5	18	14
ABS/PC	2.6	4.2	7	8
Modified nylon	1.5	3.6	4	7
PBT/PET	1.0	1.6	3	3
Modified PBT	0.7	1.3	2	2
PC/PBT	0.5	1.0	2	2
PC/PET	0.5	1.0	2	2
PSF/PET	0.3	0.9	1	1
Modified PET	0.1	0.2	—	—
Total	35.7	54.7	100	100

Table 5-24 Estimated Consumption of Specialty Polymer Blends and Alloys in the Power Tools Industry, 1987

Product	Million lb	$ Million	% of total Pounds	Dollars
Modified nylon	2.0	4.8	40	53
ABS/PC	1.9	2.4	38	27
ABS/SMA	0.4	0.5	8	6
PBT/PET	0.3	0.5	6	6
PC/PBT	0.2	0.4	4	4
Modified PET	0.1	0.2	2	2
PC/PET	0.1	0.2	2	2
Total	5.0	9.0	100	100

Table 5-25 Estimated Consumption of Specialty Polymer Blends and Alloys in the Transportation Industry, 1987

Blend/alloy	Million lb	$ Million	% of total Pounds	Dollars
PVC/acrylic	5.9	11.8	34	41
PPO/PS	5.4	8.2	31	28
SAN/EPDM	3.0	5.0	17	17
ABS/PVC	1.2	1.4	7	5
PP/EPDM	1.1	1.1	6	4
Modified ASA	0.5	0.8	3	3
Modified acetal	0.1	0.2	1	1
PPO/nylon	0.1	0.2	1	1
Modified nylon	[a]	[a]	—	—
PC/PE	[a]	[a]	—	—
Total	17.3	28.7	100	100

[a] Minor

Table 5-26 Estimated U.S. Consumption of Specialty Polymer Blends and Alloys in the Transportation Industry by Application, 1987

	Million lb									
Application	PVC/ acrylic	PPO/ PS	SAN/ EPDM	ABS/ PVC	PP/ EPDM	Modified ASA	PPO/ nylon	Modified acetal	Total	% of total
Trucks										
Interior	1.1	4.0	0.1	1.2	—	—	—	—	6.4	37
Recreational vehicles	—	—	2.9	—	—	0.3	—	—	3.2	19
Underhood	—	—	—	—	1.1	—	—	—	1.1	6
Body	—	—	—	—	—	0.2	0.1	—	0.3	2
Total	1.1	4.0	3.0	1.2	1.1	0.5	0.1	—	11.0	64
Aerospace										
Interior	4.8	[a]	—	—	—	—	—	[a]	4.8	28
Exterior	[a]	[a]	—	—	—	—	—	—	—	—
Total	4.8	—	—	—	—	—	—	—	4.8	28
Marine	[a]	1.4	[b]	—	—	[a]	—	[a]	1.5	8
Total	5.9	5.4	3.0	1.2	1.1	0.5	0.1	0.1	17.3	100

[a] Minor
[b] Quantity not known.

Table 5-27 Estimated Consumption of Specialty Polymer Blends and Alloys in the Recreation Industry, 1987

Blends and alloys	Million lb	$ Million	% of total Pounds	% of total Dollars
Modified nylon	3.6	8.7	78	78
PC/PE	0.8	2.0	18	18
PC/PBT	0.1	0.2	2	2
PC/PET	0.1	0.2	2	2
Modified ionomer	[a]	[a]	[a]	[a]
Total	4.6	11.1	100	100

[a] Minor

Table 5-28 Estimated Consumption of Specialty Polymer Blends and Alloys in the Medical Industry, 1987

Blends and alloys	Million lb	$ Million	% of total Pounds	% of total Dollars
PPO/PS	5.6	8.5	95	96
ABS/PC	0.3	0.4	5	4
Total	5.9	8.9	100	100

Table 5-29 Estimated Sales of Specialiy Polymer Blends and Alloys by Major Supplier, 1987

Supplier	Million lb	$ Million	% of total Pounds	% of total Dollars	Cumulative Percent Pounds	Cumulative Percent Dollars
General Electric	236.1	380.5	43	46	43	46
Du Pont	32.4	74.3	6	9	49	55
Borg-Warner	38.4	53.3	7	6	55	61
Mobay	24.7	39.8	4	5	60	66
Dexter/RPI	37.5	38.3	7	5	67	70
Monsanto	27.5	34.2	5	4	72	74
Himont	31.4	32.0	6	4	77	78
Dow	17.8	28.6	3	3	81	82
A. Schulman	19.3	20.9	3	3	84	84
Hoechst Celanese	10.6	19.9	2	2	86	87
Rohm & Haas	8.4	16.8	2	2	87	89
Teknor Apex	5.8	5.8	1	1	89	89
Allied	1.8	4.3	—	1	89	90
Wilson-Fiberfil	1.8	4.3	—	1	89	90
Ferro	4.0	4.2	1	1	90	91
Thermofil	3.4	4.0	1	—	91	91
Arco	3.0	3.9	1	—	91	92
LNP	1.7	3.4	—	—	91	92
Amoco	1.2	3.4	—	—	92	93
NYCOA	1.3	3.1	—	—	92	93
Emser	0.8	0.9	—	—	92	93
BASF	—	—	—	—	92	93
Others[b]	44.3	56.3	8	7	100	100
Total	553.2	833.2	100	100		

6 *APPLICATIONS OF POLYMERS IN AUTOMOTIVE PARTS*

The use of polymers in automotive parts is among the largest end-use markets. This chapter contains a compilation of useful data and application descriptions presented in tabular and drawing form. (See Figures 6-1 through 6-33 and Tables 6-1 through 6-12.)

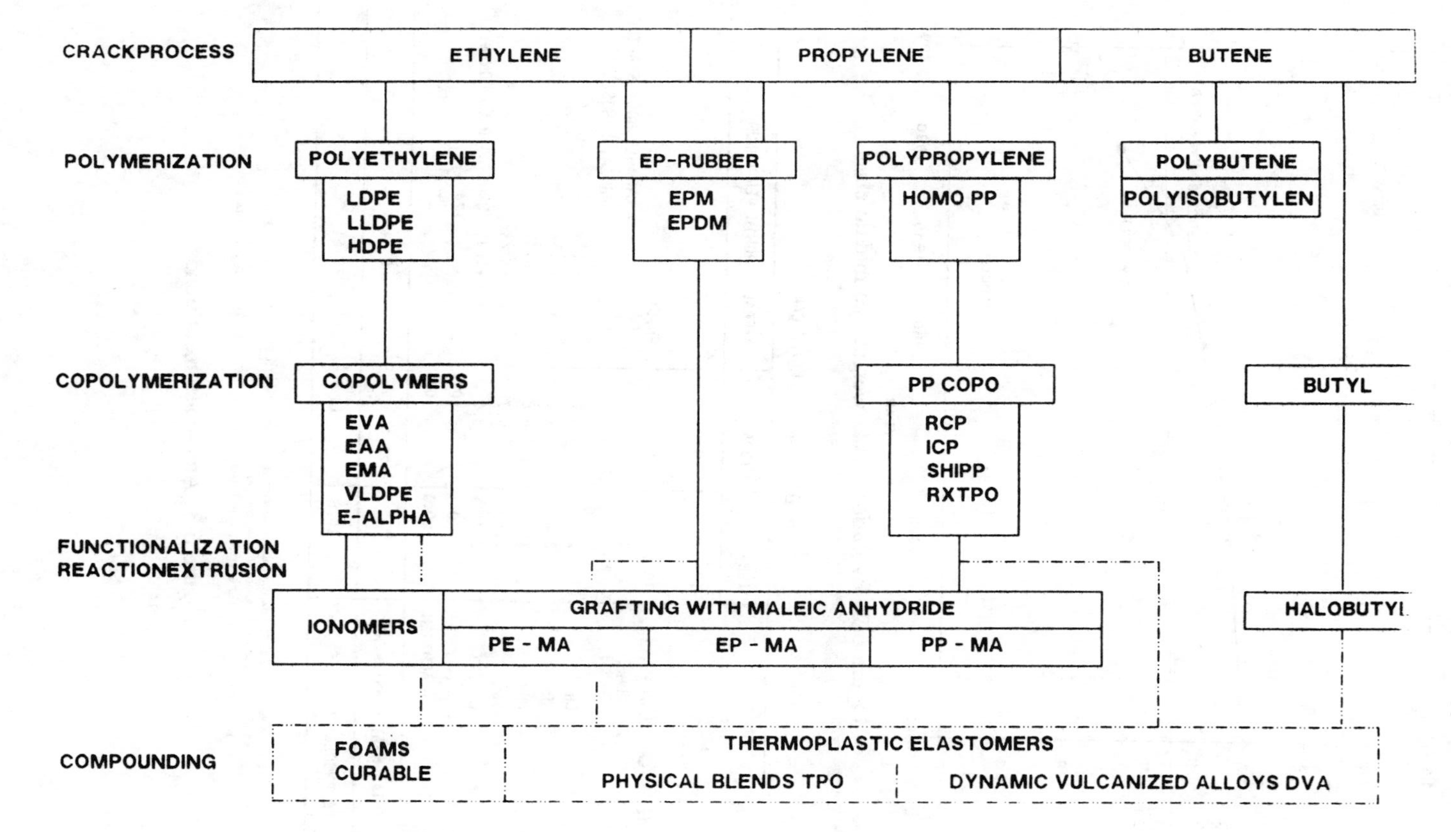

Figure 6-1 Material concept of polyolefin supplier

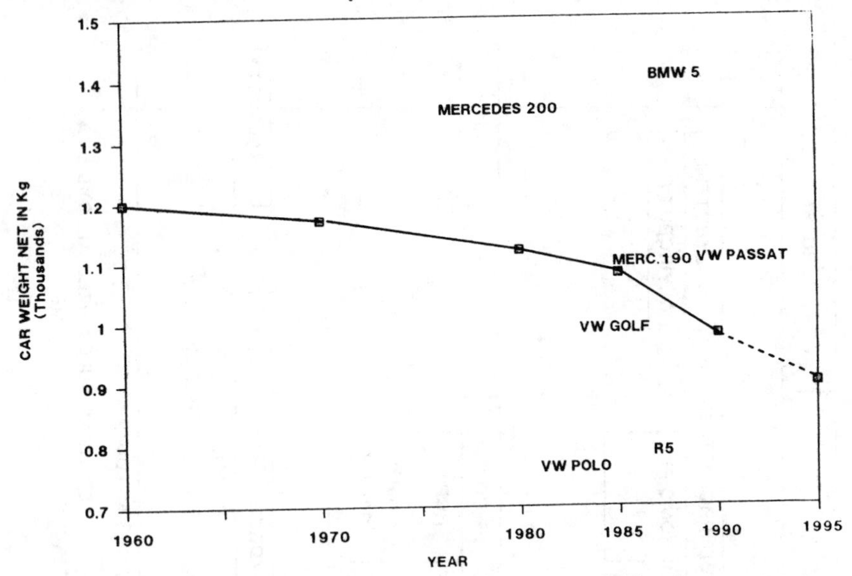

Figure 6-2 Evolution of car weight—pc middle class

WESTERN EUROPE, KG/PC CAR

	1985	1995	AVERAGE GROWTH PER ANNUM
TIRES	10	10	
TECHNICAL COMPONENTS	13.7	12.9	-6.5%
(*) EPDM	4.1	5.5	+3% : - DYNAMIC BODY SEALING - RADIATOR HOSES
CR	1.4	0.8	-5.4%
NR	2.1	1.8	-1.5%
NBR	1.9	0.8	-8.3%
SBR	2.5	0.7	-1.2%
OTHERS	1.7	3.3	+6.9% (UNDERHOOD COMPONENTS)
- ECO	0.1	0.9	+24.6
- ACM/AEM	0.1	0.5	+17.5
TOTAL NEAT ELASTOMERS	23.7	22.9	-0.3%
PC PRODUCTION (EEC), MM	10.8	13.9	+2.6%

(*) NOTE : AS A RULE OF THUMB, TO OBTAIN TOTAL EPDM CONSUMPTION IN SEGMENT, PROCEED AS FOLLOWS :
- MULTIPLY UNIT WEIGHT BY NUMBER OF PASSENGER CARS (= A)
- MULTIPLY A BY 1.4. THIS FACTOR ACCOUNTS FOR COMMERCIAL VEHICLES, REPLACEMENT MARKET AND PRODUCTION SCRAP.

Figure 6-3 Auto-elastomers trends

WESTERN EUROPE, Kg/CAR

	1985	1995	Average Growth Per Annum	
Thermoset Plastics	26.8	26.2	−0.2%	
P. EST	7.1	7.0	−0.1%	
PU	17.4	17.9	0.2%	
Phenolics	2.3	1.3	−5.9%	
Thermoplastics	36.7	44.2	1.9%	
PP/TPO	16.1	25.1	4.5%	Bumpers, Dashboards
HDPE/PE	2	3.6	6.1%	Fuel Tanks
PVC	15.9	12.5	−2.4%	Soft TPO, RTPO
Acrylics	1.3	1.3		
PS	1.4	1.7		
ENG. Thermoplastics	19.8	33.8	5.5%	Metal Replacement
ABS	8.7	7.5	−1.5%	Stiff TPO, EPP
PA	5.6	12.5	8.4%	
PC	1.4	4.2	11.6%	
POM	1.8	2.9	4.9%	
PPO	1.4	2.5	5.9%	
PET/PBT	0.4	3.3	23.5%	Electrical Components/ Blends With PC
Total Plastics	83.3	104.2		

Figure 6-4 Autoplastics trends

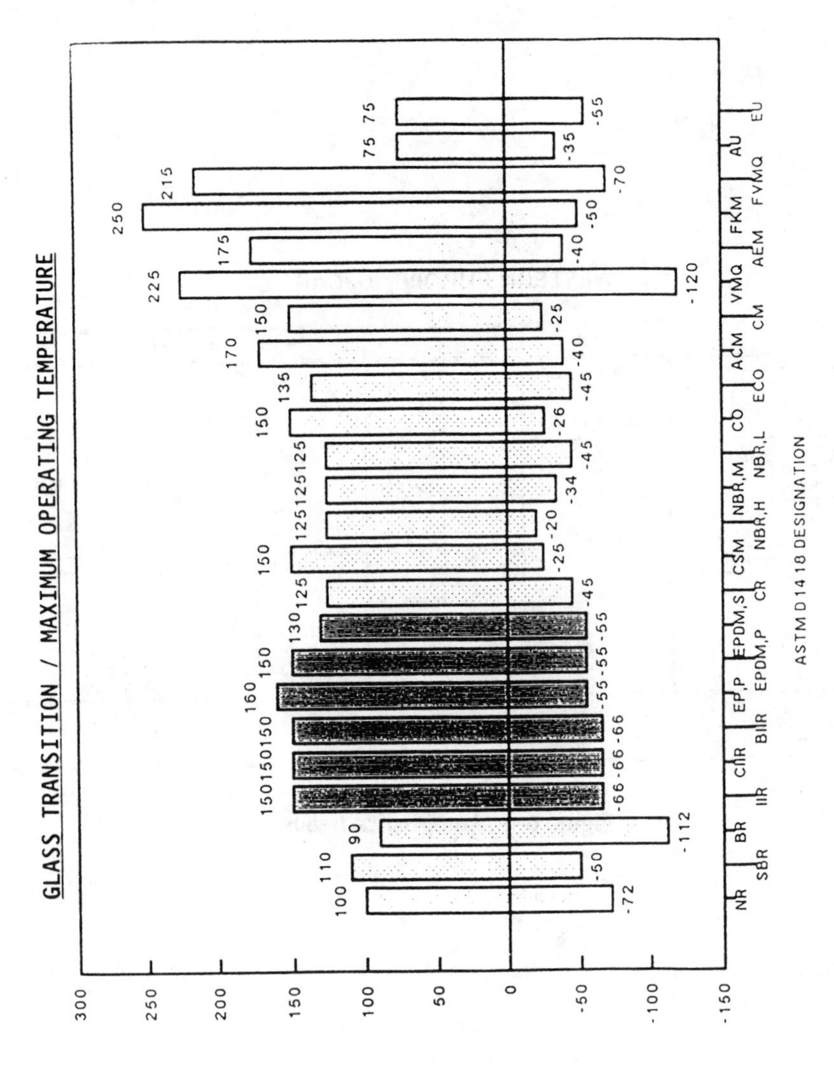

Figure 6-5 Elastomer service ranges

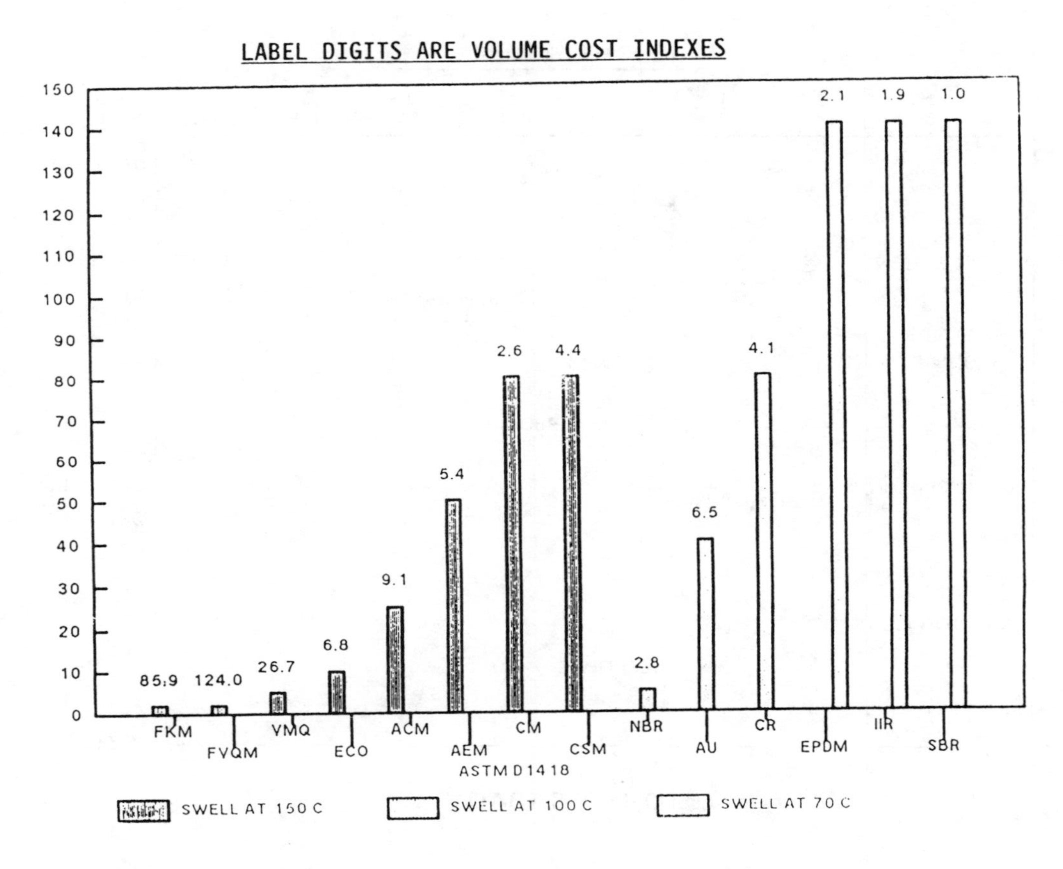

Figure 6-6 Elastomers oil resistance-cost balance

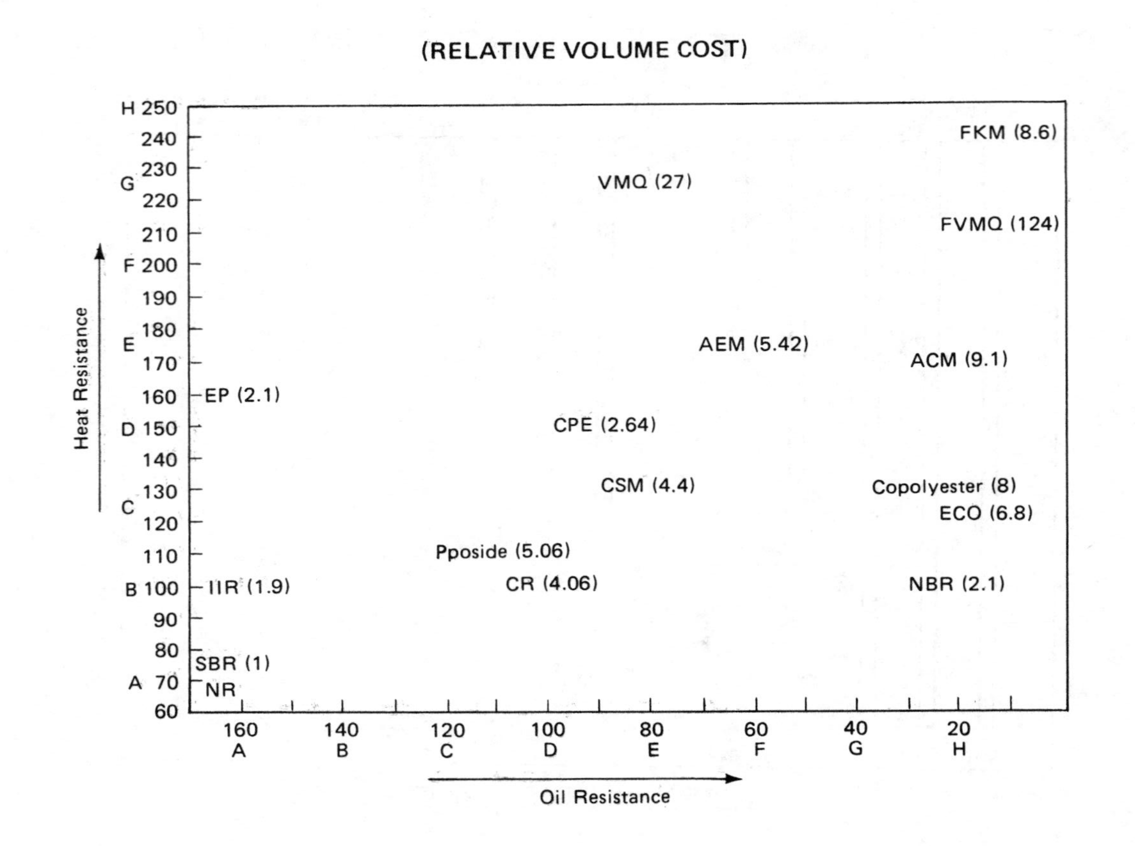

Figure 6-7 Polymers performance (ASTM D2000)

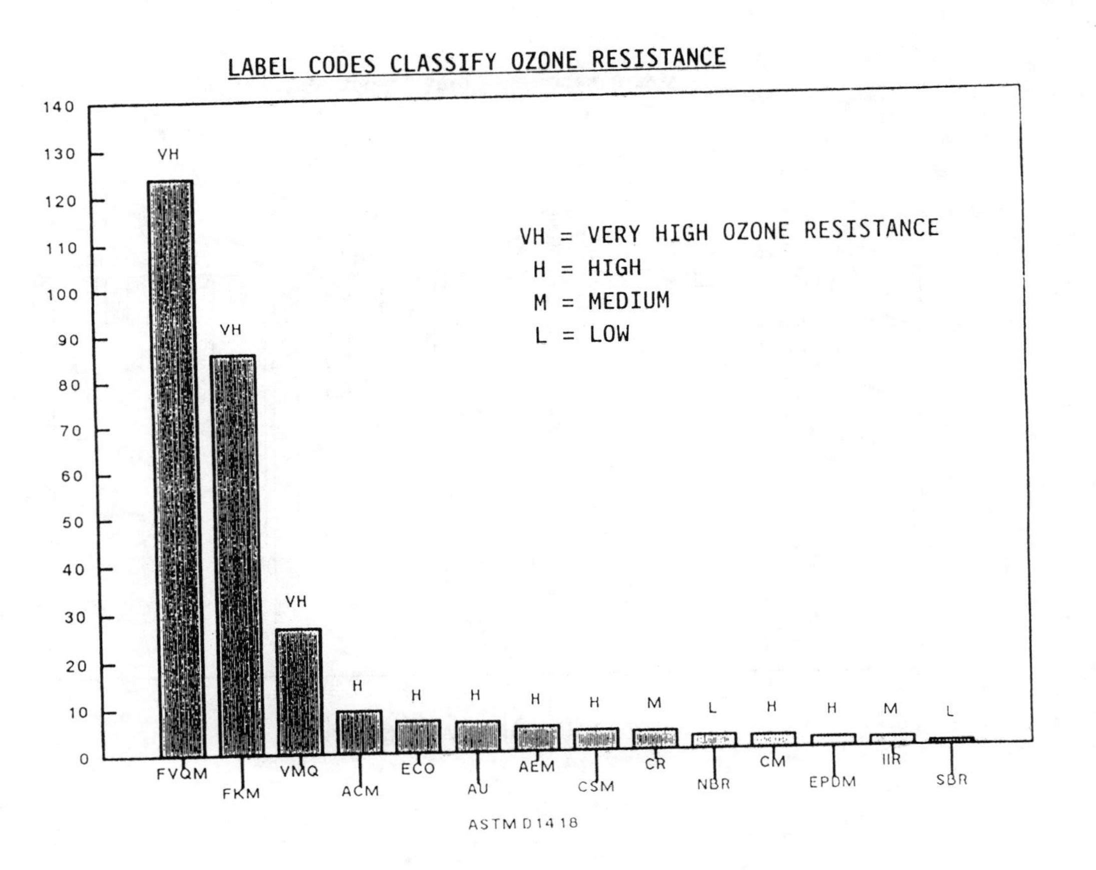

Figure 6-8 Volume cost and ozone resistance map

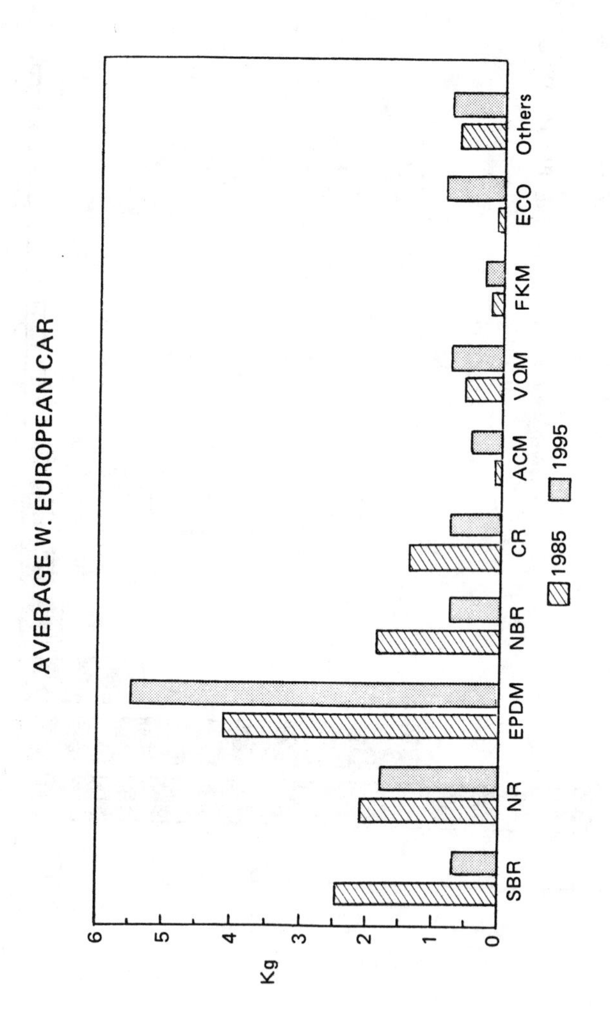

Figure 6-9 Auto - elastomer trends

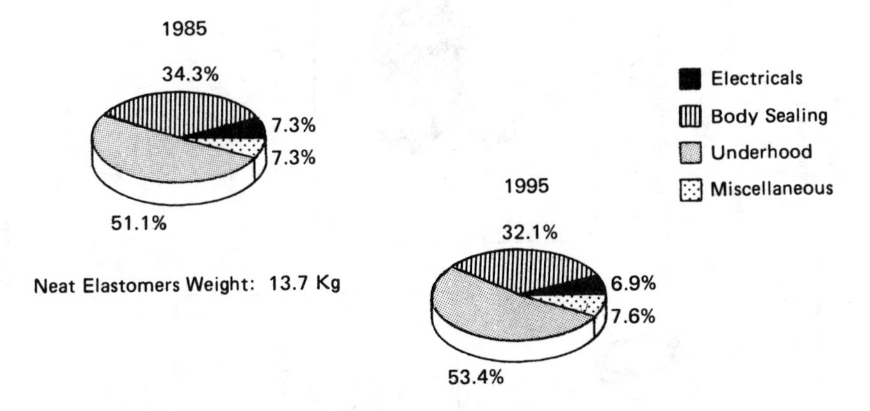

Figure 6-10 Neat elastomers by application category

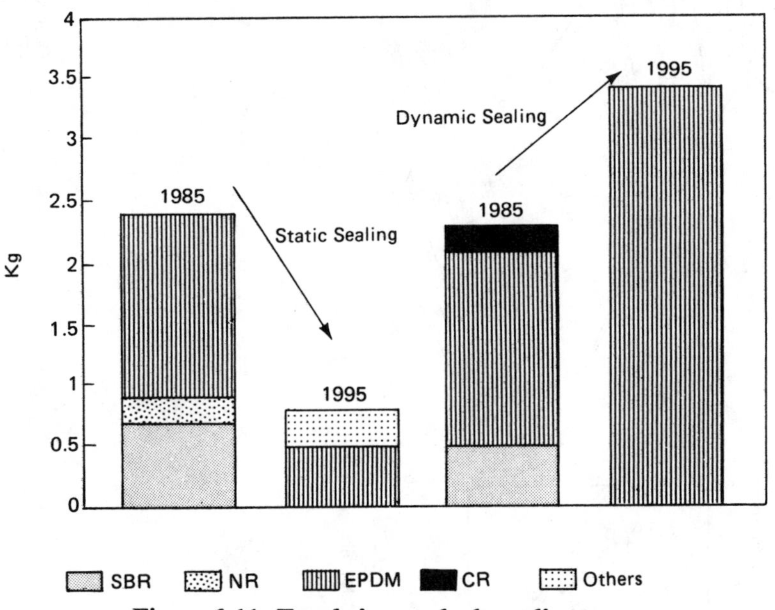

Figure 6-11 Trends in auto body sealing

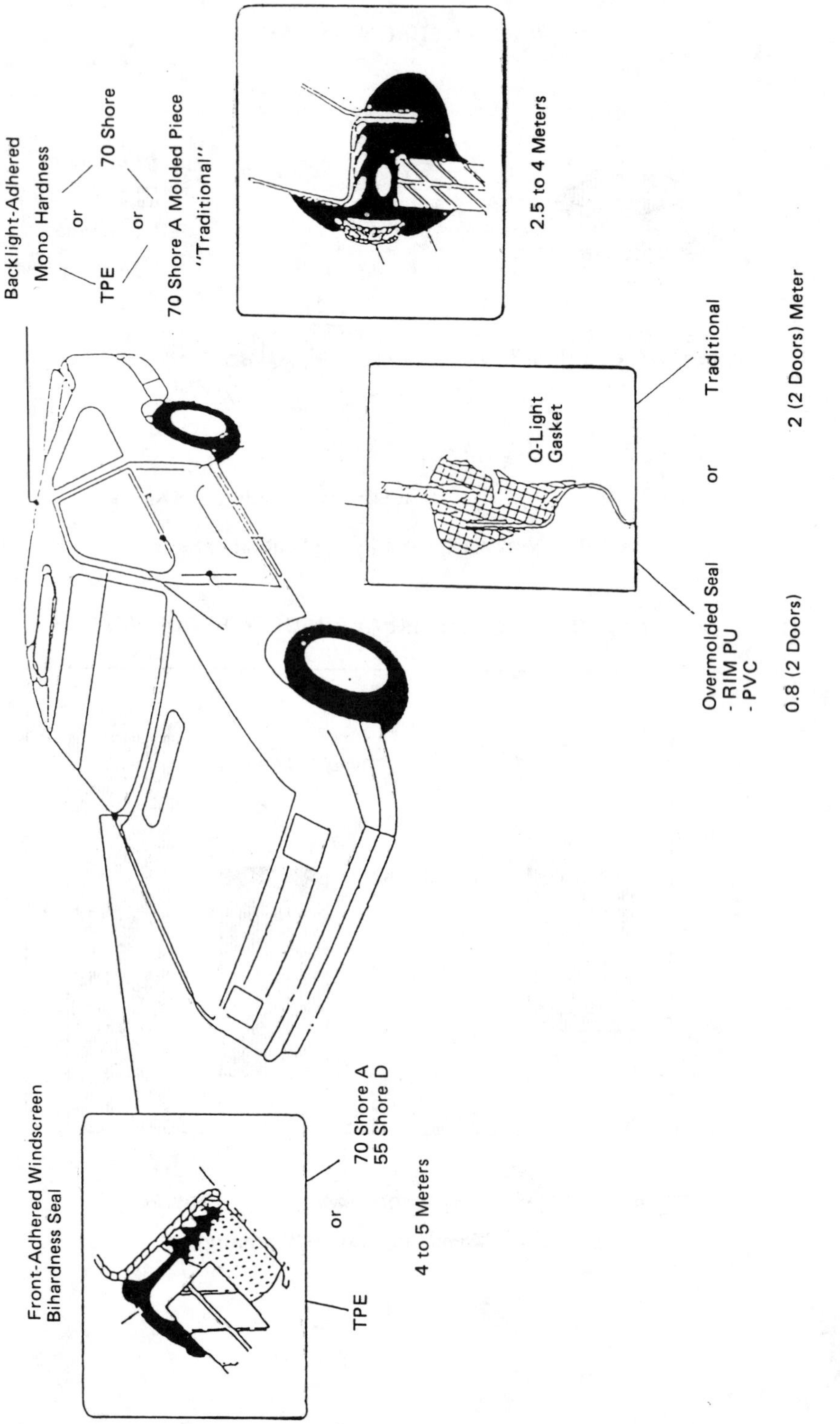

Figure 6-12 Static window sealing

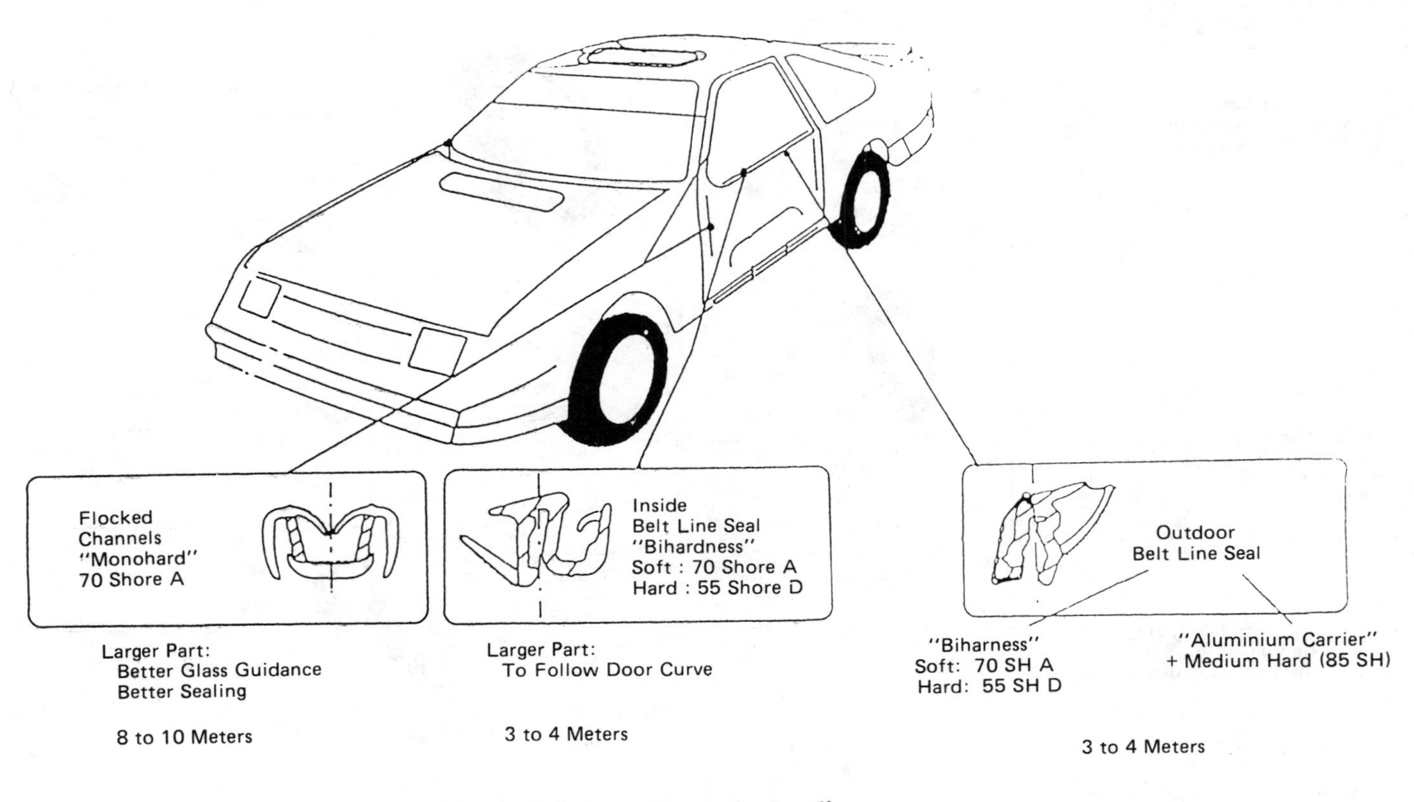

Figure 6-13 Dynamic window sealing

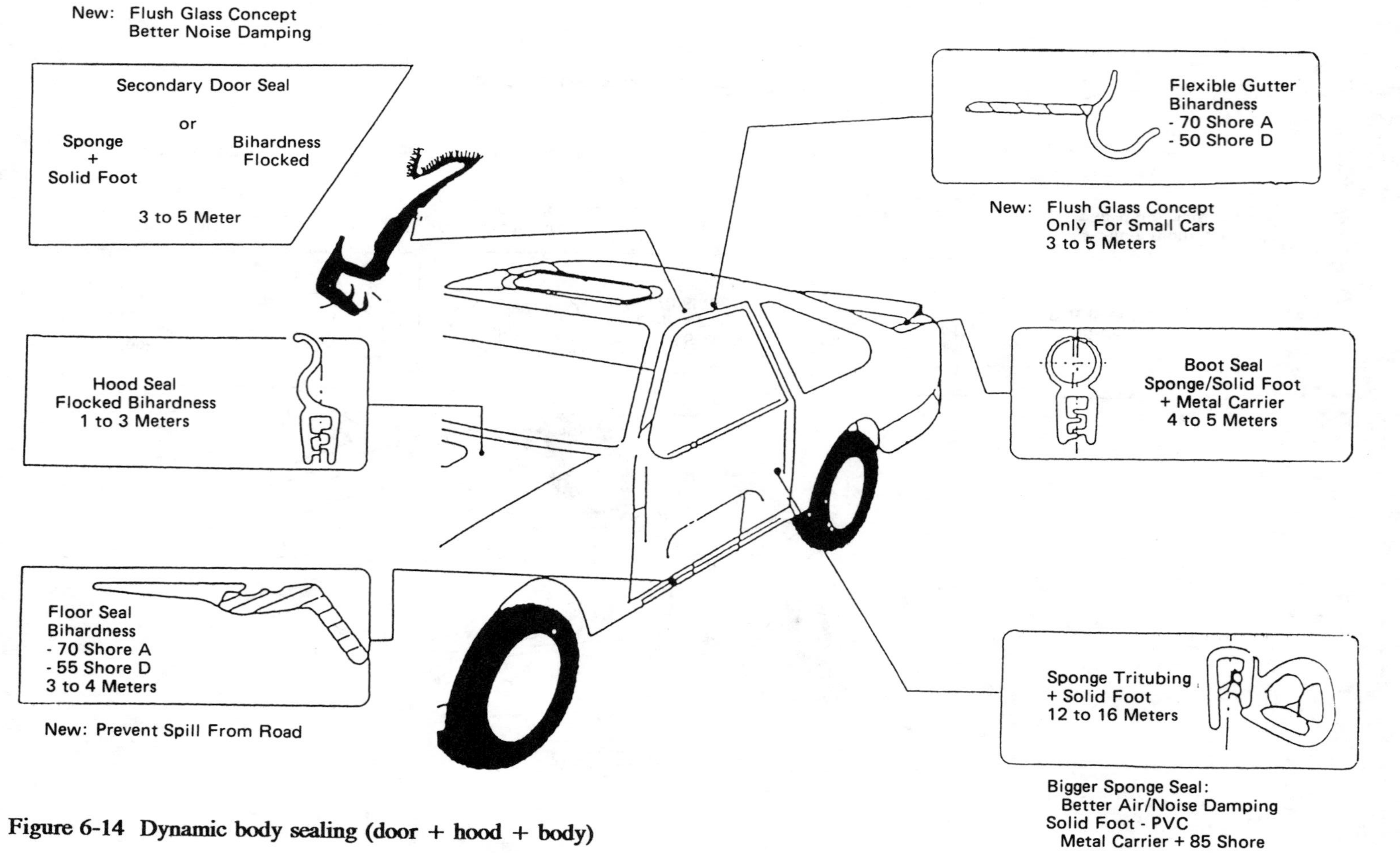

Figure 6-14 Dynamic body sealing (door + hood + body)

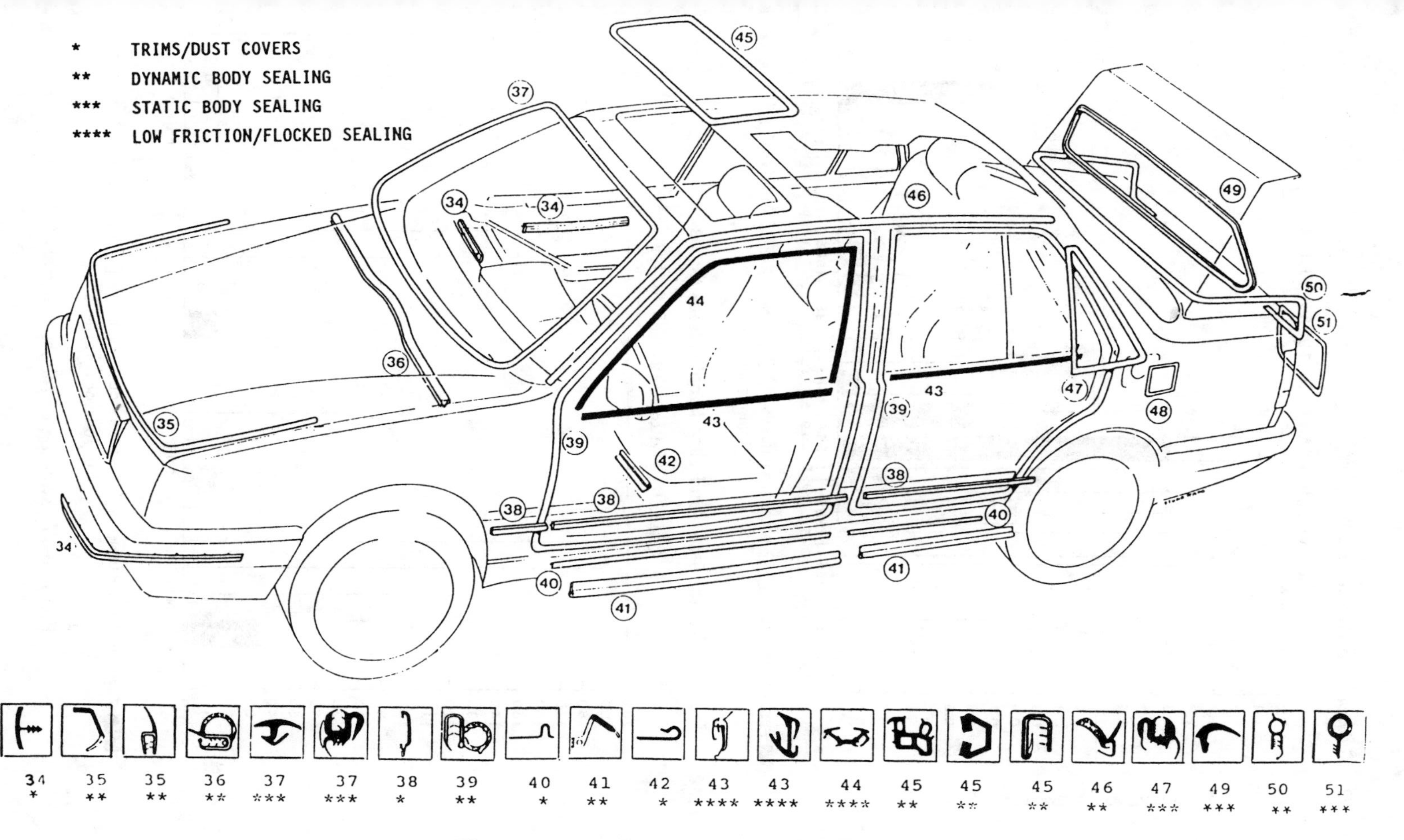

Figure 6-15 Body sealing overview - West Europe

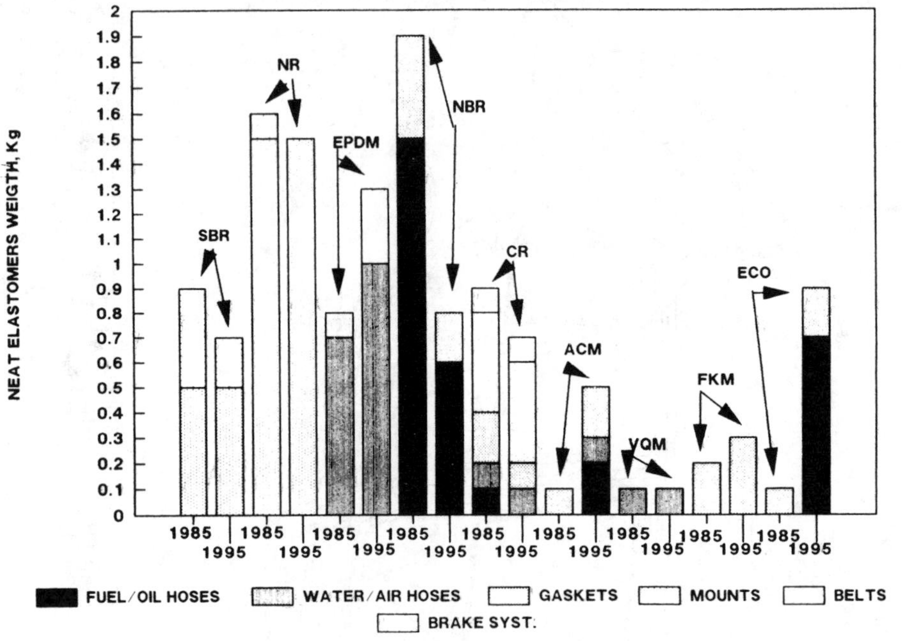

Figure 6-16 Underhood applications trends

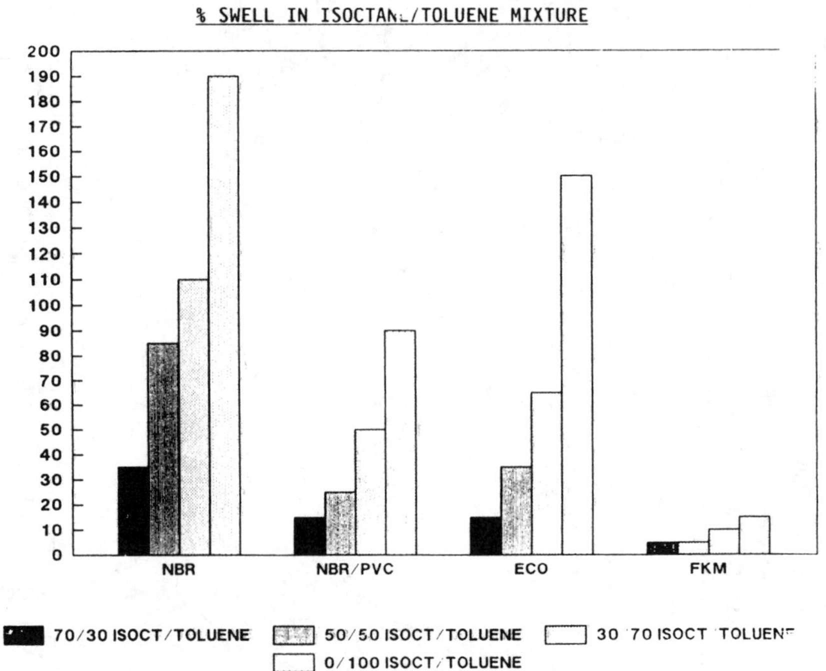

Figure 6-17 Swelling of rubber materials

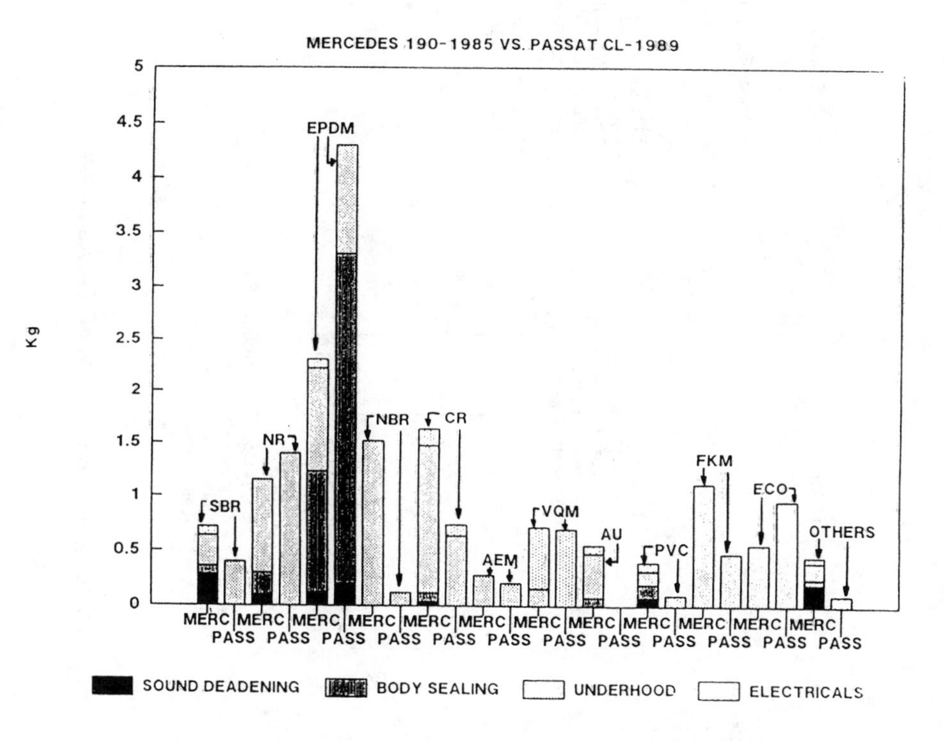

Figure 6-18 Neat elastomers usage in passenger cars

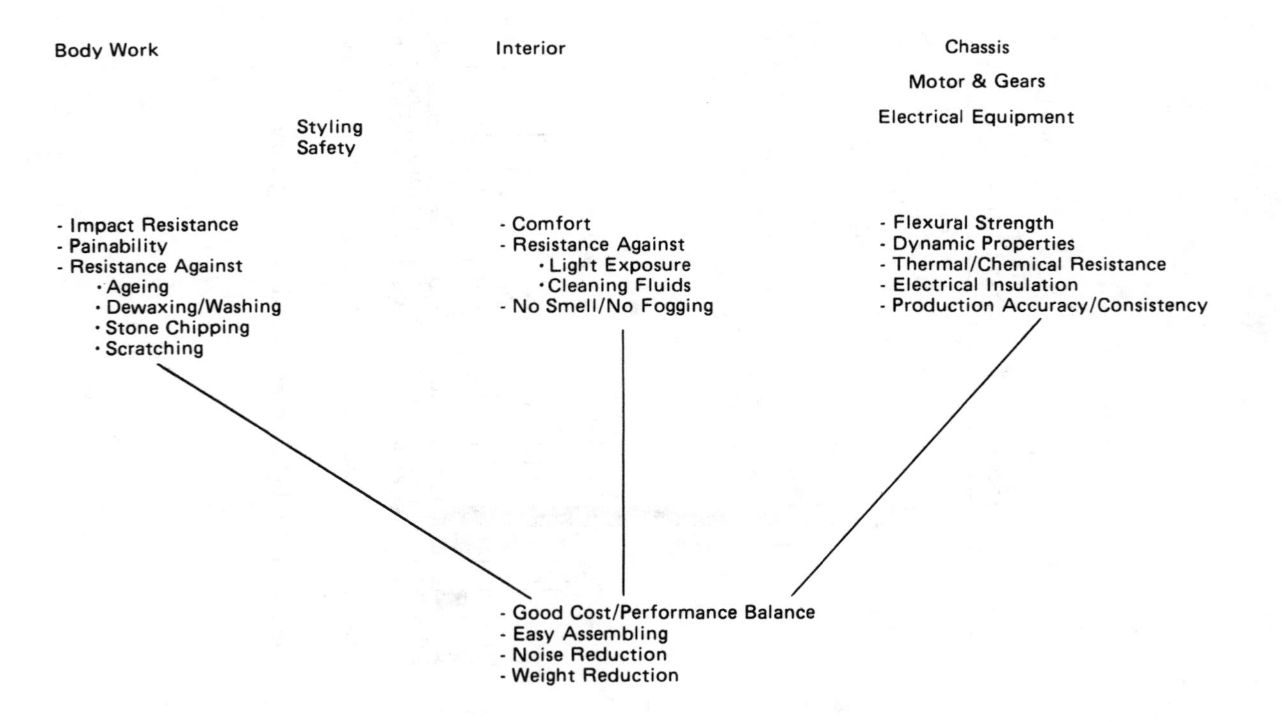

Figure 6-19 Basic requirements of plastics in auto

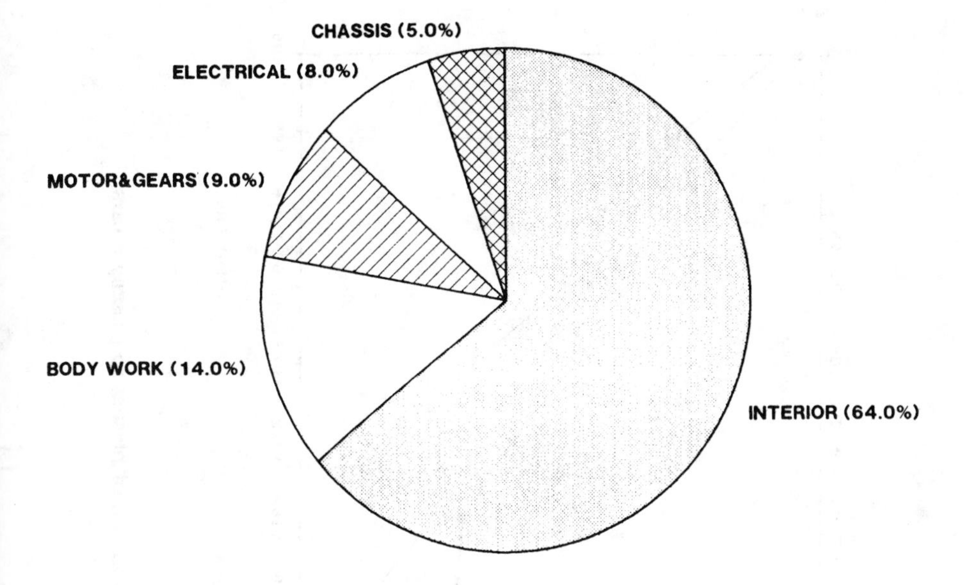

Figure 6-20 Plastics range of applications

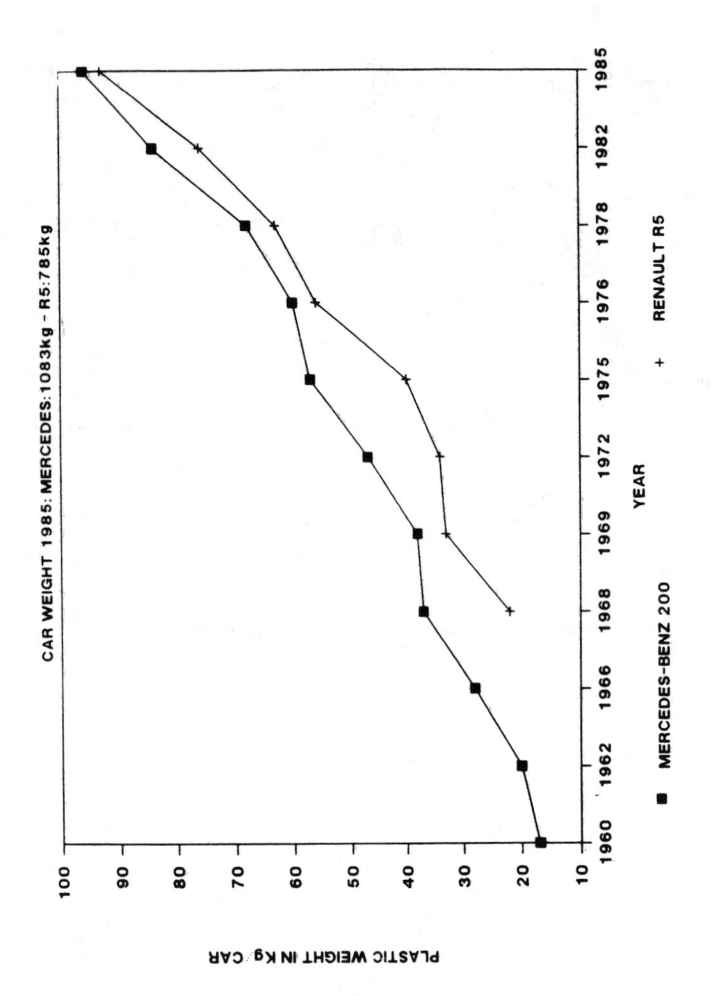

Figure 6-21 Evolution of plastics in passenger cars

RENAULT 5
MERCEDES 200

	Renault 5 (1972)	Mercedes 200 (1976)	Renault Sup 5 (1985)	Mercedes 200 (1985)
Car Weight, Kg	730	1350	785	1250
Plastic Weight, Kg	32.68	58.1	92.64	98.2
PE	0.33	2.6	9.72	4.2
PP/TPO	1.49	9.4	17.38	20.6
ABS/PS	1.37	10.8	7.6	8.5
PPO Blends	0.24	-	0.82	-
PMMA	0.78	1.7	0.79	1.1
PC/PC Blends	0.19	0.3	1.79	5.7
POM	0.51	0.5	0.61	4.0
PA	1.41	6.0	5.37	17.3
PVC	10.9	14.2	12.16	14.9
Cellulosics	0.72	-	2.3	-
P Est	-	0.3	3.16	7.1
PU	5.06	10.5	13.80	14.0
Phonolics	2.07	1.8	0.97	0.8
SMC	7.61		16.17	

Figure 6-22 Evolution of plastic usage

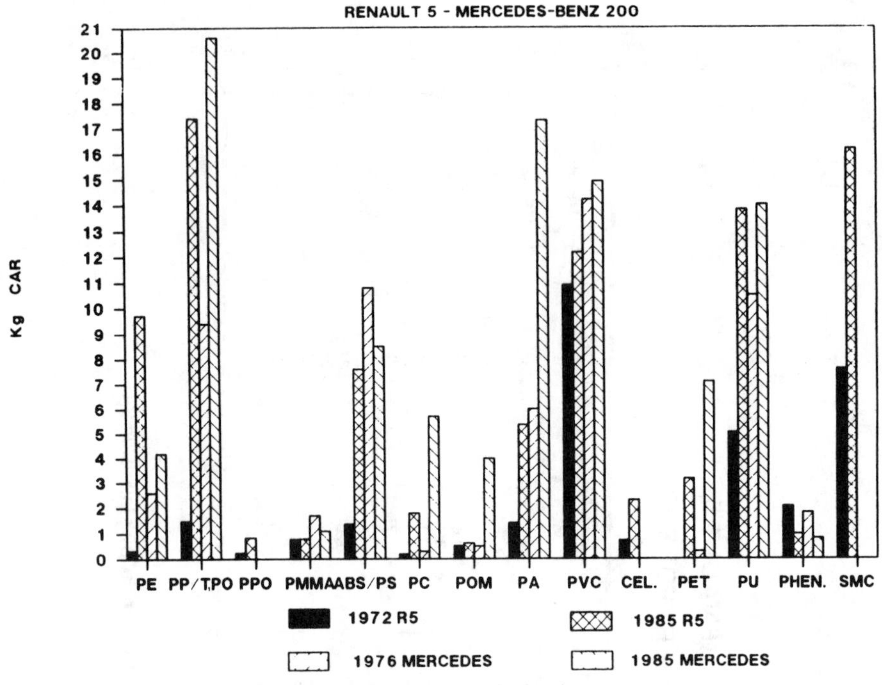

Figure 6-23 Evolution of plastic usage

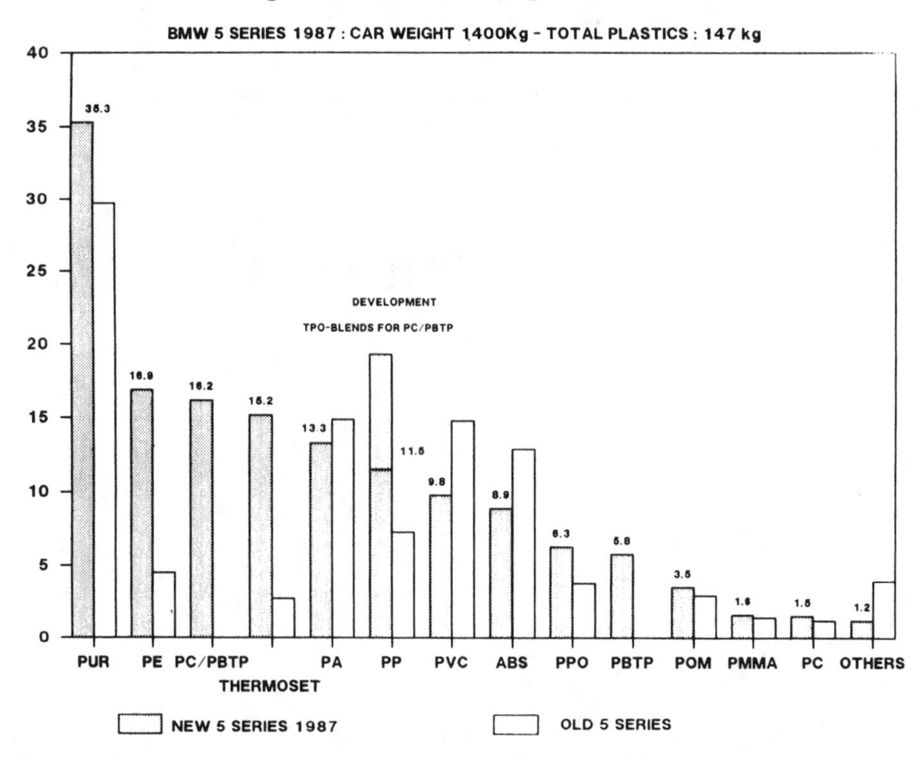

Figure 6-24 Evolution of plastic usage

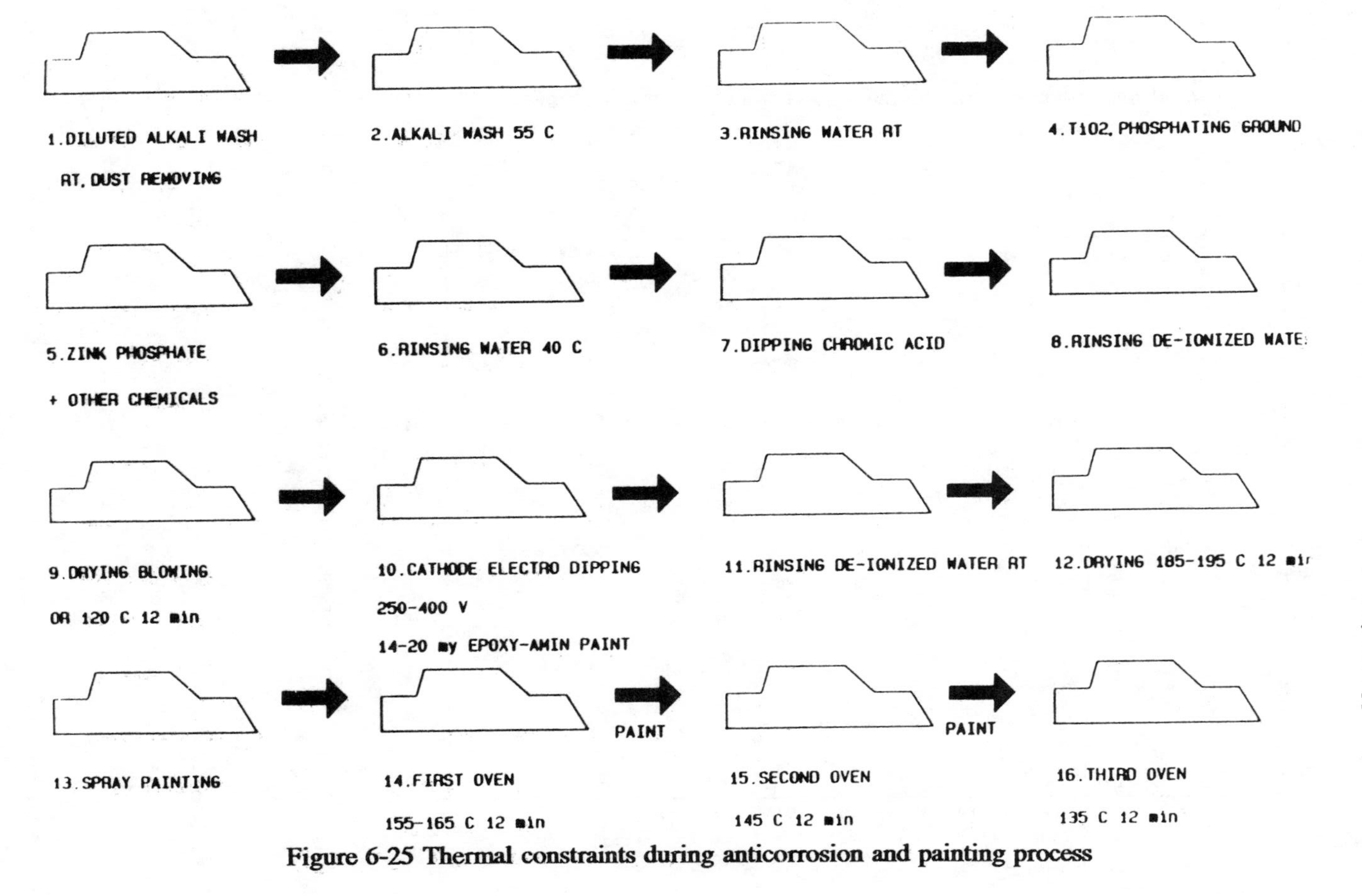

Figure 6-25 Thermal constraints during anticorrosion and painting process

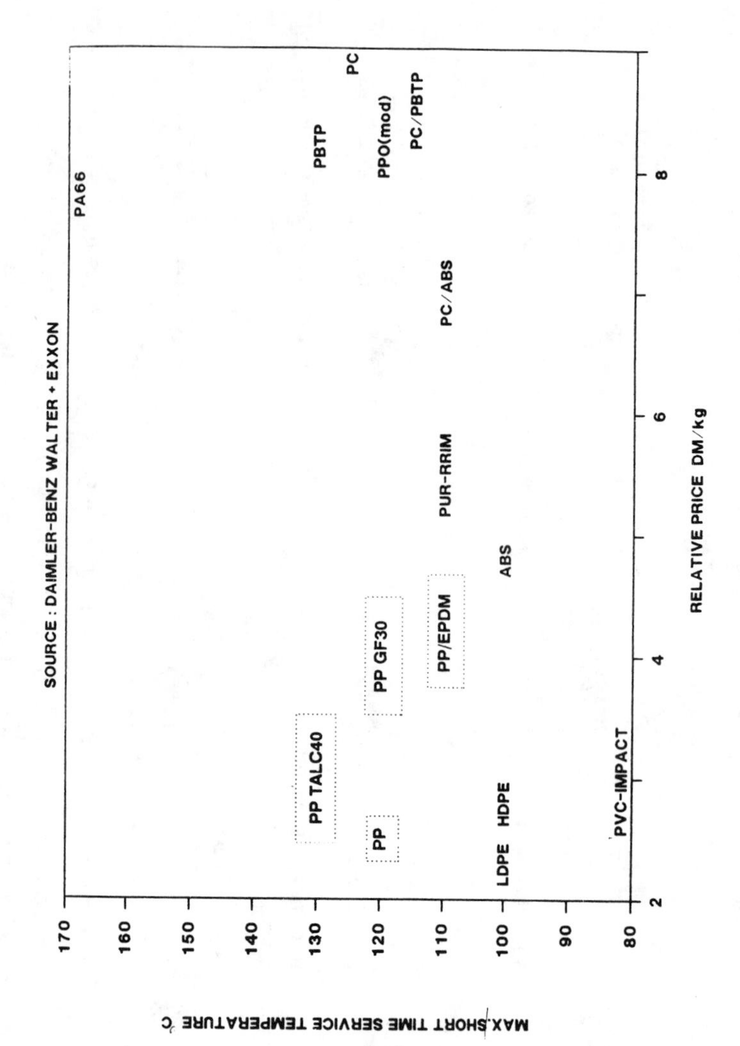

Figure 6-26 Maximum service temperature versus material price

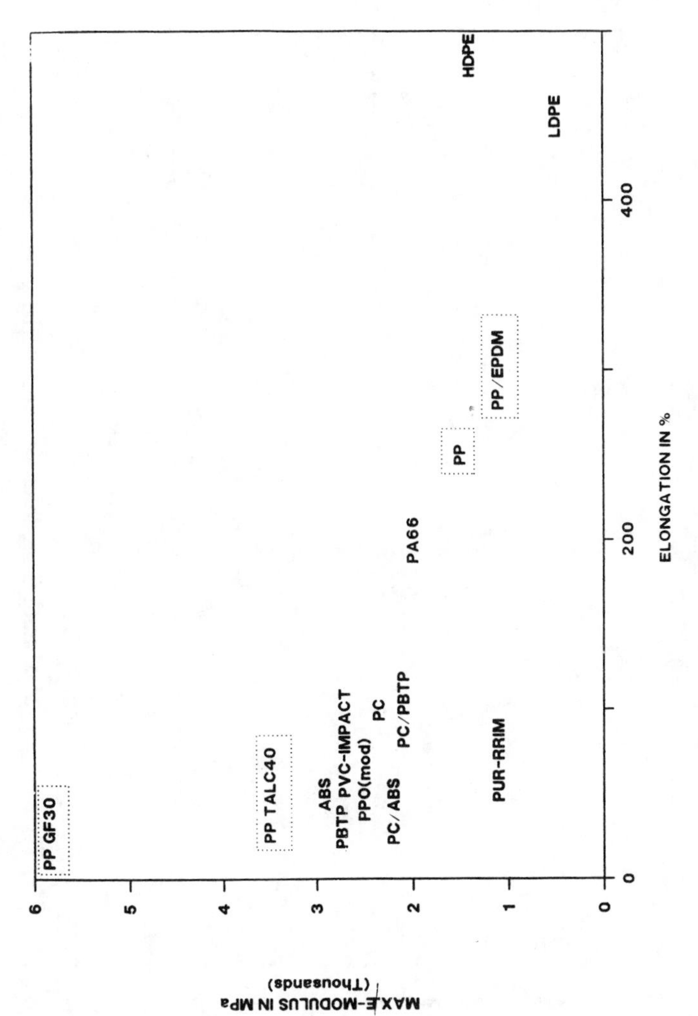

Figure 6-27 Maximum E-modulus versus elongation

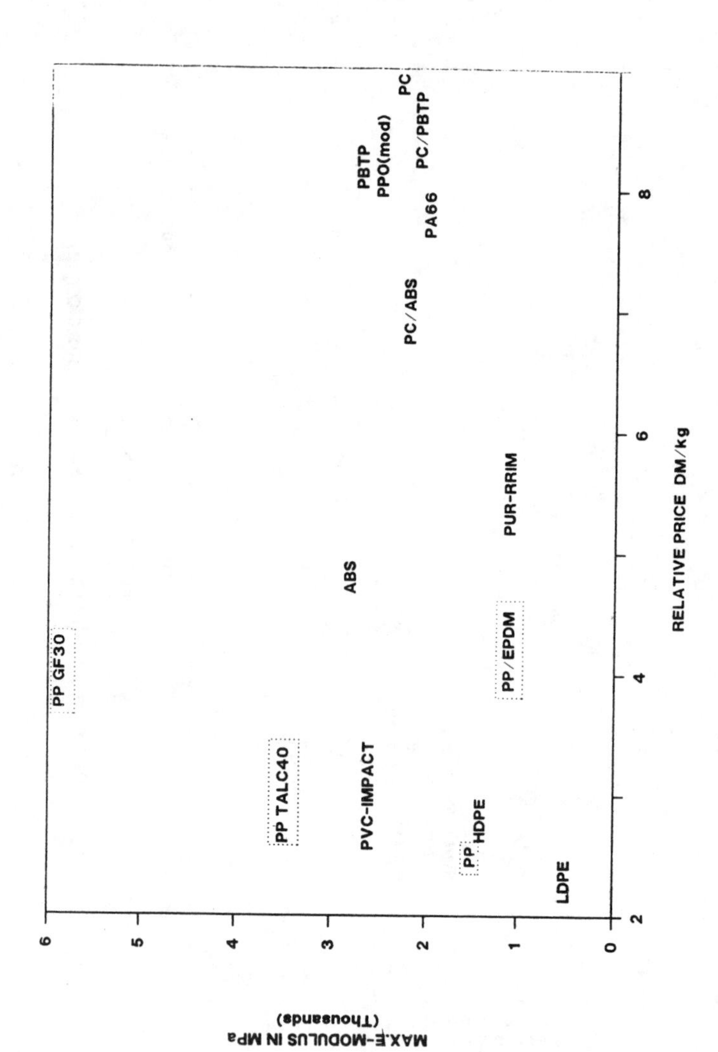

Figure 6-28 Maximum E-modulus versus material price

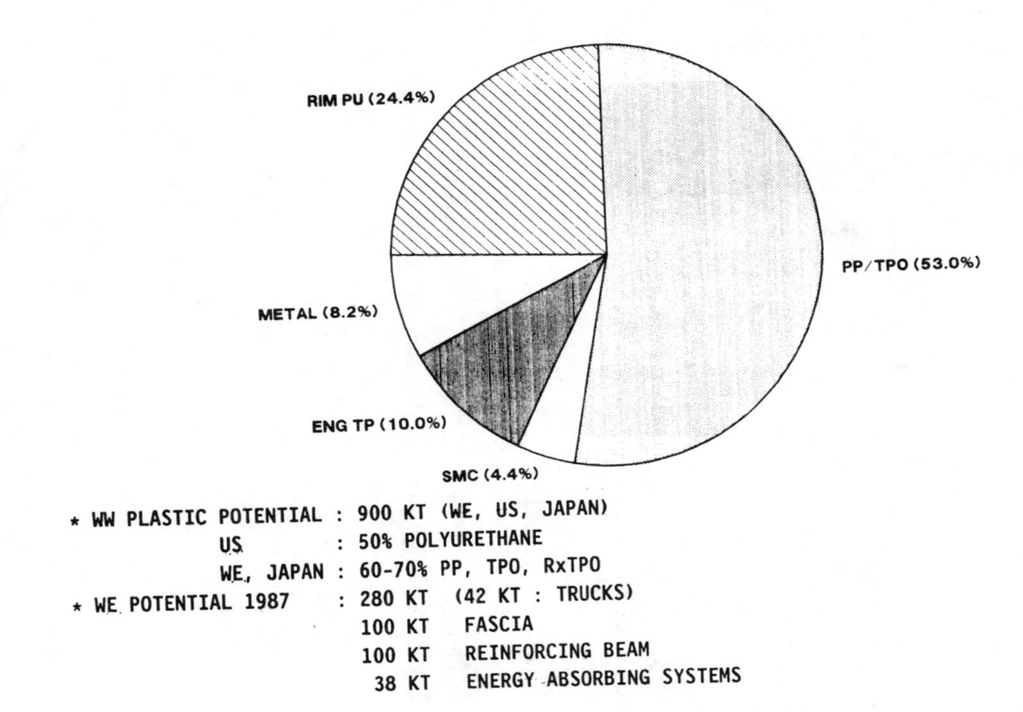

Figure 6-29 World pc car bumper fascia market - 1987

Material	Raw Material Cost kg/DM (%)	Material Thickness (mm)	Specific Weight g/cm3	Total Volume Based on 17,500 cm² Outside Area cm3	Weight Per Pair (kg)	Material Cost DM
PC	9.0 (100)	2.5	1.22	4375	5.3	47.25
PP	4.0 (44.5)	.5	0,90	8750	7.9	31.50
SMC	5.5 (61)	2.5	1.8	4375	7.9	43.30
RRIM	7.0 (78)	3.0	1.2	5250	6.3	44.10

Material	Injection Molding Clamping Force, T	Cycle Time Min	Production Cost Index	Total Cost Per Car, % Ideal Case - No Scrap	Total Cost Per Car, % Shorter Cycle Time + Scrap
PC	2700	1.5	1	64	83
PP	2750	2.5	1.8	77	52
SMC	800	3.0	1.7	100	100
RRIM	50	2.5	1.6	95	95

- PP longer cycle time due to thickness
New: 3.5 - 4.5 mm
PC actual: 3 0 4 mm

- Scrap rates PC: 3%
PP: <1%
- PP recycling direct
- same part

Figure 6-30 Specific cost comparison—plastic bumper materials

Present Materials ABS, PC, PC/ABS, PA-GF, PP

Trends

- Grilles Painted in Body Color
ABS ===> Modified PPO Painting on Line
- Borders Chromium Plated
- Suppression of Grilles and Integration of Air Inlets in Bumpers

Bonnet Air Grilles

Metal ===> Modified PPO Dimensional Stability (Large Parts)
Some ABS (For Small Parts)

Exterior Trims

Materials: Modified PPO, RIM-PU
TPE (Under Study)
Filled PP Less Suitable (Low-Temperature Resistance)

Wheel Covers

Materials: Modified PPO, PA

Figure 6-31 Radiator grilles

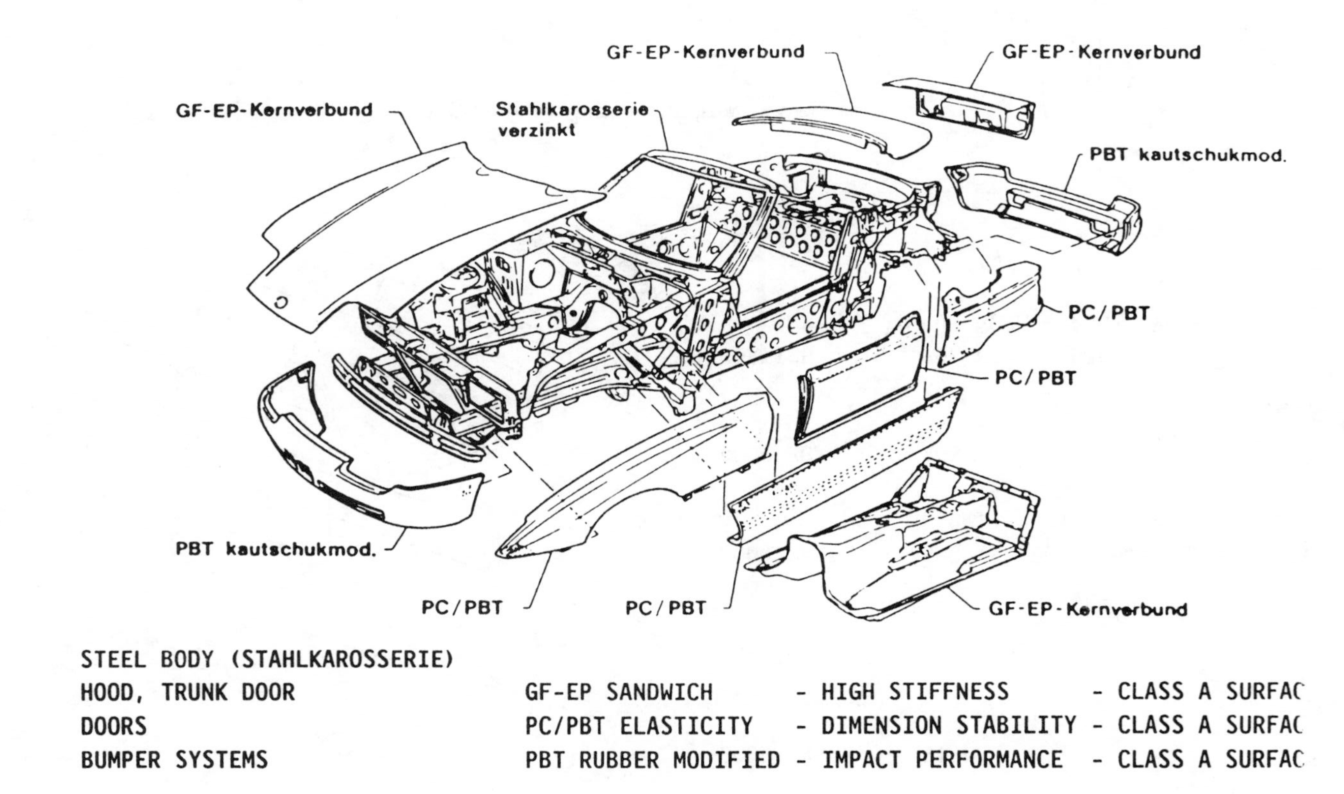

Figure 6-32 BMW Z1 roadster bodywork

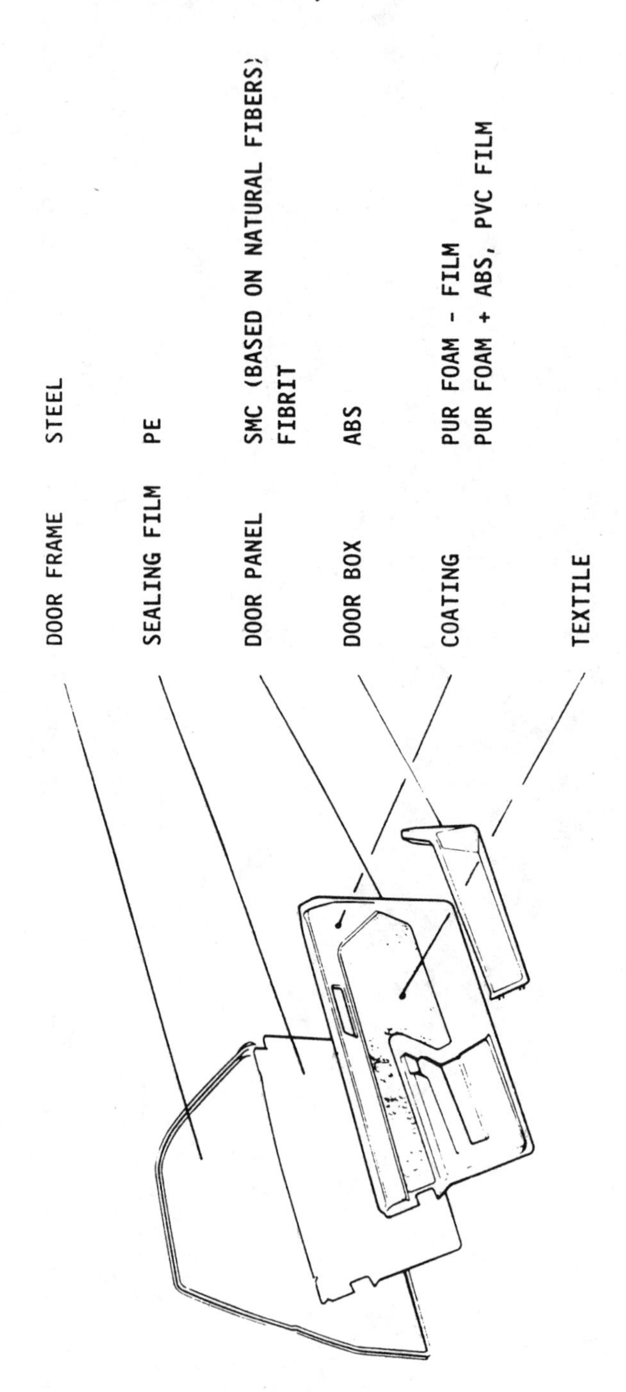

Figure 6-33 Design internal door panel

Table 6-1 Automotive Component Classification

Soft < 400 MPa	Semistiff 400-1,000 MPa	Stiff 1,000-3,000 MPa	Structural > 3,000 MPa
Applications by Stiffness Range (Flexural Modulus)			
Body sealing trims/	Bumper fascias	Dasboards	Parcel shelves
gaskets/weatherstripping	Splash shields	Wing panels/	Floor panels
Bellows	Wheelhouse liners	body panels	Fan support front panels
suspension	Mud flaps	Steering wheels	Bumper beams
steering	Spoilers	Consoles	Engine noise shield
air duct	Fuel tank	Air filters	Battery tray
Ignition wires	Ski carriers	Glove boxes	Front arm rest
Grommets		Baby seats	Door panels
Drain tubes		Instrument panels	Lateral protection panels
Dust cover caps		Wheel trims	Seat structures
Shock absorber pads for		Grilles	
suspension system/mounts		Mirror housing	
Adhesive films		Headlamp housing	
upholstery/panels		Battery cases	
seats		Rear seat panels	
Noise insulation			
gap-filling foams			
Hoses/belts			
Bumper shock absorbers			
Water Reserviors			

Table 6-1 Automotive Component Classification (*Continued*)

Soft < 400 MPa	**Semistiff 400-1,000 MPa**	**Stiff 1,000-3,000 MPa**	**Structural > 3,000 MPa**
Key Functional Requirements			
Elastic sealing/ flex fatigue/ precision application/ processing latitude	High-impact performance/ surface finishing	Self-support/high Dimensional stability/ surface finish (if exposed) or compatibility to other substrates	Load-bearing/metal replacement
Type of Materials			
EPR CIIR, IIR TPE Soft TPO SP.PE, PE	TPO (unfilled) ICP HDPE	PP (filled: talc, mica whiting, GF) EPP	STC

Table 6-2 Elastomer Overview

Elastomer	ASTM D1418		ASTM D2000	Price per liter $	SG
Fluorosilicone	FVMQ		FK	124	141
Fluorocarbon	FKM	VITON	HK	85.93	1.86
Silicone	VMQ	VINYL-SILOXANE	FC-FE-GE	26.7	1.38
Polyacrilate	ACM		DF-DH	9.07	1.13
Epichlorohydrin	ECO		CH	6.83	1.30
Polyurethane	AU		BG	6.53	1.12
Ethylene Acrylic	AEM	VAMAC	EF	5.42	1.12
Chlorosulphonated PE	CSM		CE	4.39	1.28
Polychloropene	CR		BC-BE	4.06	1.23
Nitrile rubber	NBR, H		BF-BJ-BK	2.81	1
Chl PE	CM		BE-CE	2.64	1.25
Ethylene propylene	EPDM, S		AA-BA-CA	2.1	0.86
Butyl rubber	IIR		AA-BA	1.86	0.92
Styrene-butadiene	SBR		AA-BA	1	1.01
PP oxide		PAREL	BC	5.06	1
Thermoplastic polyester		HYTREL	CH	8.05	1.2

Table 6-3 Sealing Component Overview

Component type	Technology change 1980s to 1990s	Systems/Material
Static Window Seal		
Front Window Seal backlite Q-light gasket	Adhered window (bihardness concept) Adhered window + direct seal overmolding	Smaller profiles (one third weight) Rim PU or PVC in competition with x-linked rubber
Dynamic Window Seal		
Glass channel	Deeper/larger part to improve glass guidance	Traditionally, flocked thermoset 70 shore A. Low friction coating at introductory stage (PA film/HDPE coating).
Belt line seal indoor	Larger part to follow the door curved shape	Bihardness technology stiff 55 shore D (snap-on) soft 70 shore A
Belt line seal outdoor	Minimize rubber as visual part thin seal for better drag coefficient	Aluminium carrier (light-rigid-nonrusty) Coextruded with 85 shore A cross-linked rubber

Table 6-3 Sealing Component Overview (*Continued*)

Component type	Technology change 1980s to 1990s	Systems/Material
Dynamic Door Sealing		
Sponge door	Tritubing to improve water tightness/air damping	Rubber attributes 50% bigger, but low load deflection in range -30°C to +100°C low friction/smooth surface
Floor seal	New - to prevent spill from the road	Bihardness technology 55 shore D and 70 shore A
Secondary door seal combination with flexible gutter	New - part of flush glass technology with improved air damping/water tightness	Bihardness technology or sponge/solid foot + metal carrier
Hood Seal	New trend to bihardness with flocked soft part more seal under hood (aerodynamic)	55 shore D 70 shore A
Buot Seal	No change	Sponge/solid foot metal carrier

Table 6-4 Automotive Hoses — Development Trends

	1985 Materials	Development Trends	Opportunities
Radiator/Heater	EPDM	EP - Peroxide	New grade with best processability/property balance
Fuel system	NBR	ECO	–
	NBR + PVC	PA-11/ECO	–
		FKM/AEM CSM	EP alloying (to improve processability of AEM/CSM)
Power steering	NBR/CR	AEM or CM	EP alloying (to improve processability of AEM/CSM)
Air conditioning	NBR	CSM NBR/EPDM	Functionalized EP (to improve processability of CSM or to bond EPDM cover to NBR liner)
Oil cooling	NBR	AEM	EP alloying (to improve processability of AEM)
Brake hoses	SBR/CR	EPDM	Conversion underway

Table 6-5 Mercedes 190 – 1985 Model

	Neat Elastomers Content, Kg												
	SBR	NR	BR	EPDM	NBR	CR	ACM	Q	PU	PVC	FKM	ECO	Others
Tires (5)	6.2	0.4	3.1	0	0	0	0	0	0	0	0	0	0
Mats	0.28	0.1	0	0.12	0	0.04	0	0	0	0.08	0	0	0.2
Profiles	0.08	0.2	0	1.11	0	0.08	0	0	0.08	0.12	0	0	0.05
Radiator Hoses	0	0	0	0.75	0	0	0	0	0	0	0	0	0
Fuel Hoses	0	0	0	0	0.48	0.28	0	0	0	0	0.16	0.08	0
Oil Hoses	0	0	0	0	0.44	0.16	0	0	0	0	0.16	0.08	0
Other Hoses	0.08	0	0	0.03	0.2	0.16	0.14	0.08	0	0.4	0.08	0.24	0.1
Gaskets	0	0	0	0.06	0.2	0.04	0.14	0.08	0.08	0	0.4	0.08	0
Bushes	0.08	0.65	0	0.06	0.04	0.04	0	0	0.08	0	0	0	0
Mounts	0.08	0.2	0	0.03	0.04	0.16	0	0	0.24	0	0	0	0
V-Belts	0	0	0	0	0	0.2	0	0	0	0	0	0	0
Wiring	0.08	0	0	0.09	0	0.16	0	0.56	0.08	0.08	0	0	0.05
Bellows	0	0	0	0.06	0.08	0.32	0	0	0	0.08	0.16	0	0
Membranes	0.04	0	0	0	0.04	0	0	0	0	0	0.16	0.08	0.05
Total	6.92	1.55	3.1	2.31	1.52	1.64	0.28	0.72	0.56	0.4	1.12	0.56	0.45

Table 6-6 PP Opportunities Summary

Interior Parts	Plastic penetration very high Replacement of expensive material by cheaper thermoplastics Trend favored by material properties upgrading and/or advances in fabrication process
PP-based materials expected to become more competitive	

New Technologies	Key Effects
Reactor modification	Lower cost
High crystallinity	Higher-temperature performance
Filler compounding know-how	Lower cost/tailored
Glass fiber reinforcement	Performance increase
STC technology	Broader application
Alloys/blends with EP	Ranges

Table 6-7 ETP Service Temperature

Exterior Parts	Temperature, °C	Materials
For resistance to main part line	170-190	Noryl GTX 900 (Mod P) SMC
For resistance to paint retouch stage	135-155	PC/PBT PPE/PA ABS/PA
Interior Parts		
For resistance to paint retouch stage	115-125	PPO/HIPS PC/ABS PC/ASA PPE/HIPS ABS/SMA
For interior service only	95-105	ASA ABS

Table 6-8 Thermal Limits of Various Plastics

		Baking Temperature - Current Lines		
	Thermal Limits °C	190/180 °C ED Primer	140/130 °C Paint Baking (In Line)	100/80 °C Paint Baking (Off Line)
Reinforced PBT	150	–	Y	Y
PBT	140	–	Y	Y
Reinforced PA	142	–	Y	Y
PA	120	–	–	Y
Reinforced PC	128	–	–	Y
PC	125	–	–	Y
Reinforced PP (Long GF)	120	–	–	Y
Reinforced PP (Short GF)	102	–	–	Y
PP	90	–	–	Y (Marginal)
R-RIM PU	120	–	–	Y
RIM PU	105	–	–	Y
BMC	220	Y	Y	Y
SMC	220	Y	Y	Y

Y = Yes

Table 6-9 Polymer Blends — Heat Distortion Temperature above 100°C

Material	Tradename	Supplier
ABS/PC	BAYBLEND PROLOY, RONFALIN, TERBLEND B	Bayer Borg-Warner, DSM, BASF
ASA/PC	TERBLEND S	BASF
SMSA/PC	ARLOY	Arco
PPE/HIPS	NORYL, LURANYL, PREVEX, XYRON	General Electric, BASF, Borg-Warner, ASAHI
PPE/PA	NORYL GTX, ULTRANYL, MPPE	General Electric, BASF, ASAHI
ABS/PA	ELLMID, TRIAX	Borg-Warner, Monsanto
PC/PBT	XENOY MAKOBLEND ULTRABLEND	General Electric Bayer BASF
PBTP/ASA/GF	ULTRABLEND S	BASF
PBTP/PETP	VALOX	General Electric
POLYACRYLATE/PETP	ROPET	Roehm & Haas
POLYSULFONE/ABS	MINDEL	Union Carbide
POLYSULFONE/PETP/GF	MINDEL	Union Carbide

Table 6-10 Material Requirements for Interior Parts

Good performance characteristics	Abrasion resistance Easy cleaning Resistant against cleaning fluids
Production in large scale	Punching/transform ability Welding/adhesion/sewing
Variations in color/decor/grain/gloss	
Coating ability	Painting Printing Stamping Plating
Compatability to different materials Thermal/weathering/dimension stability Flame resistance U.S.-stanndard FMVSS 302 Low fogging High functionality (instrument panel, seats) Easy assembling Low weight	

Table 6-11 Support Materials Interior

Product	Production / Materials / Processing	Supplier
FIBRIT	Wooden fibers in aqueous phase with binder (Phenolic resin) seperated (wet feed) Compression molding - drying/compressing - steam activation	FIBRIT, KREFELD + licence
LIGNOTOCK	Wooden fibers with phenolic resin solution steam activation compression (only in direction)	PAG, ESSEN SONTRA, FULDA
FIBR MAT	Wood and other cellulose fibers in dry process Bound with resins (2%-20%) - no steam activation thermoforming	Van Dresser, USA
CEKAZELL	Wooden and textile fibers - Lignotock principle	Kast, Gernsbach
PELIFORM	SMC - semifinished product (natural fibers with resins) hot-mold forming	EMPE, GERTSRIED
TRIFLEX	Nonwoven based on cotton, phenolic resin thermoforming	Borgers, Bocholt
CEEFAPUR	Glass fiber nonwoven support - form foamed - PUR system	Freudenburg, Weinheim
HM-PP	Wood fiber filled PP Compression/thermoforming	Elastogran, Lemfoerde
PP-GM	Glasmat reinforced PP Flow forming/thermoforming	Symalit, Lenzburg, Elastogran

Table 6-12 Internal Panels

Class	Material	
Low/Medium	PP	
Medium/high	RIM PU or Lignotock	PVC coated
High	RIM PU or Lignotock	Foamed with PU and PVC coated

	RIM PU	versus	Lignotock
Material cost	High		Low
Tooling cost	Low		High

Trends	PVC coating	Decreasing
	Textile coating	Increasing
	PP support with soft painting	Increasing

APPENDIX A

ABR	Acrylate-butadiene rubber
ABS	Acrylonitrile-butadiene-styrene rubber
ACM	Acrylate rubber
AES	Acrylonitrile-ethylene-propylene-styrene quater-polymer
AMMA	Acrylonitrile-methyl methacrylate copolymer
ANM	Acrylonitrile-acrylate rubber
APP	Atactic polypropylene
ASA	Acrylonitrile-styrene-acrylate terpolymer
BIIR	Brominated isobutene-isoprene (butyl) rubber
BR	Cis- 1,4-butadiene rubber (cis- 1,4-polybutadiene)
BS	Butadiene-styrene copolymer (see also SB)
CA	Cellulose acetate
CAB	Cellulose acetate-butyrate
CAP	Cellulose acetate-propionate
CF	Cresol-formaldehyde resin
CHC	Epichlorohydrin-ethylene oxide rubber
CHR	Epichlorohydrin rubber (see also CO)
CMC	Carboxymethyl cellulose
CN	Cellulose nitrate (see also NC)
CNR	Carboxynitroso rubber; (tetrafluoroethylene-tri-fluoronitrosomethane-unsat.monomer terpolymer)
CO	Poly[(chloromethyl)oxirane]; epichlorohydrin rubber (see also CHR)
CP	Cellulose propionate
CPE	Chlorinated polyethylene
CR	Chloroprene rubber
CS	Casein

CSM	Chlorosulfonated polyethylene
CTA	Cellulose triacetate
CTFE	Poly(chlorotnfluoroethylene); (see also PCTFE)
EAA	Ethylene-acrylic acid copolymer
EVA	Ethylene-vinyl acetate copolymer
EC	Ethyl cellulose
ECB	Ethylene copolymer blends with bitumen
ECTFE	Ethylene-chlorotrifluoroethylene copolymer
EEA	Ethylene-ethyl acrylate copolymer
EMA	Ethylene-methacrylic acid copolymer or ethylene-maleic anhydride copolymer
EP	Epoxy resin
E/P	Ethylene-propylene copolymer (see also EPM,EPR)
EPDM	Ethylene-propylene-nonconjugated diene terpolymer (see also EPT)
EPE	Epoxy resin ester
EPM	Ethylene-propylene rubber (see also E/P, EPR)
EPR	Ethylene-propylene rubber (see also E/P, EPM)
EPS	Expanded polystyrene; polystyrene foam (see also XPS)
EPT	Ethylene-propylene-diene terpolymer (see also EPDM)
ETFE	Ethylene-tetrafluoroethylene copolymer
EVA, E/VAC	Ethylene-vinyl acetate copolymer
EVE	Ethylene-vinyl ether copolymer
FE	Fluorine-containing elastomer
FEP	Tetrafluoroethylene-hexafluoropropylene rubber; see PFEP
FF	Furan-formaldehyde resins
FPM	Vinylidene fluoride-hexafluoropropylene rubber
FSI	Fluorinated silicone rubber
GR-I	Butyl rubber (former US acronym) (see also IIR, PIBI)
GR-N	Nitrile rubber (former US acronym) (see also NBR)
GR-S	Styrene-butadiene rubber (former US acronym; see PBS, SBR)
HDPE	High-density polyethylene
HEC	Hydroxyethylcellulose
HIPS	High-impact polystyrene
HMWPE	High molecular weight polyethylene
IIR	Isobutene-isoprene rubber; butyl rubber (see also GR-I, PIBI)
IPN	Interpenetrating polymer network
IR	Synthetic cis- I .4-oolyisoprene

LDPE	Low-density polyethylene
LLDPE	Linear low density polyethylene
MABS	Methyl methacrylate-acrylonitrile-butadiene-styrene
MBS	Methyl methacrylate-butadiene-styrene terpolymer
MC	Methyl cellulose
MDPE	Medium-density polyethylene (ca. 0.93-0.94 g/cm3)
MF	Melamine-formaldehyde resin
MPF	Melamine-phenol-formaldehyde resin
NBR	Acrylonitrile-butadiene rubber; nitrile rubber; GR-I
NC	Nitrocellulose; cellulose nitrate (see also CN)
NCR	Acrylonitrile-chloroprene rubber
NIR	Acrylonitrile-isoprene rubber
NR	Natural rubber (cis- 1,4-polyisoprene)
OER	Oil extended rubber
OPR	Propylene oxide rubber
PA	Polyamide (e.g., PA 6,6 = polyamide 6,6 = nylon 6,6 in U.S. literature)
PAA	Poly(acrylic acid)
PAI	Polyamide-imide
PAMS	Poly(alpha-methylstyrene)
PAN	Polyacrylonitrile (fiber)
PARA	Poly(arylamide)
PB	Poly(1-butene)
PBI	Poly(benzimidazoles)
PBMA	Poly(n-butyl methacrylate)
PBR	Butadiene-vinyl pyridine copolymer
PBS	Butadiene-styrene copolymer (see also GR-S, SBR)
PBT, PBTP	Poly(butylene terephthalate)
PC, PCO	Polycarbonate
PCD	Poly(carbodiimide)
PCTFE	Poly(chlorotrifluoroethylene)
PDAP	Poly(diallyl phthalate)
PDMS	Poly(dimethylsiloxane)
PE	Polyethylene
PEA	Polv(ethyl acrylate)

Polymer Acronyms

PEC	Chlorinated polyethylene (see also CPE)
PEEK	Poly(arylether ketone)

PEI	Poly(ether imide)
PEO, PEOX	Poly(ethylene oxide)
PEP	Ethylene-propylene polymer (see also E/P, EPR)
PEPA	Polyether-polyamide block copolymer
PES	Polyethersulfone
PET, PETP	Poly(ethylene terephthalate)
PF	Phenol-formaldehyde resin
PFA	Perfluoroalkoxy resins
PFEP	Tetrafluoroethylene-hexafluoropropylene copolymer; FEP
PI	Polyimide
PIB	Polyisobutylene
PIBI	Isobutene-isoprene copolymer; butyl rubber; GR-I,
PIBO	Poly(isobutylene oxide)
PIP	Synthetic poly-cis- 1,4-polyisoprene; (also CPI, IR)
PIR	Polyisocyanurate
PMA	Poly(methyl acrylate)
PMI	Polymethacrylimide
PMMA	Poly(methyl methacrylate)
PMMI	Polypyromellitimide
PMP	Poly(4-methyl- I-pentene)
PO	Poly(propylene oxide); or polyolefins; or phenoxy resins
POM	Polyoxymethylene, polyformaldehyde
POP	Poly(phenylene oxide) (also PPO/PPE)
PP	Polypropylene
PPC	Chlorinated polypropylene
PPE	Poly(phenylene ether)
PPMS	Poly(para-methylstyrene)
PPO	Poly(phenylene oxide) (also PPO/PPE)
PPOX	Poly(propylene oxide)
PPS	Poly(phenylene sulfide)
PPSU	Poly(phenylene sulfone)
PPT	Poly(propylene terephthalate)
PS	Polystyrene
PSB	Styrene-butadiene rubber (see GR-S, SBR)
PSF, PSO	Polysulfone
PSU	Poly(phenylene sulfone)
PTFE	Poly(tetrafluoroethylene)
P3FE	Poly(trifluoroethylene)
PTMT	Poly(tetramethylene terephthalate) = poly(butylene terephthalate) (see also PBTP)
PUR	Polyurethane
PVA, PVAC	Poly(vinyl acetate)
PVAL	Poly(vinyl alcohol) (also PVOH)
PVB	Poly(vinyl butyral)

PVC Poly(vinyl chloride)
PVCA Vinyl chloride-vinyl acetate copolymer (also PVCAC)
PVCC Chlorinated poly(vinyl chloride)
PVDC Poly(vinylidene chloride)
PVDF Poly(vinylidene fluoride)
PVF Poly(vinyl fluoride)
PVFM Poly(vinyl formal) (also PVFO)
PVI Poly(vinyl isobutyl ether)
PVK Poly(N-vinylcarbazole)
PVP Poly(N-vinylpyrrolidone)

RF Resorcinol-formaldehyde resin

SAN Styrene-acrylonitnle copolymer
SB Styrene-butadiene copolymer
SBR Styrene-butadiene rubber (see also GR-S)
SCR Styrene-chloroprene rubber
S-EPDM Sulfonated ethylene-propylene-diene terpolymers
SHIPS Superhigh-impact polystyrene
SI Silicone resins; poly(dimethylsiloxane)
SIR Styrene-isoprene rubber
SMA Styrene-maleic anhydride copolymer
SMS Styrene-alpha-methylstyrene copolymer

TPE Thermoplastic elastomer
TPR 1,5-trans-Poly(pentenamer)
TPU Thermoplastic polyurethane
TPX Poly(methyl pentene)
UF Urea-formaldehyde resins
UHMW-PE Ultrahigh molecular weight poly(ethylene) (also UHMPE) (molecular mass over 3.1×106 g/mol)
UP Unsaturated polyester

VC/E Vinyl chloride-ethylene copolymer
VC/E/VA Vinyl chloride-ethylene-vinyl acetate copolymer
VC/MA Vinyl chloride-methyl acrylate copolymer
VC/MMA Vinyl chloride-methyl methacrylate copolymer
VC/OA Vinyl chloride-octyl acrylate
VC/VAC Vinyl chloride-vinyl acetate copolymer
VC/VDC Vinyl chloride-vinylidene chloride
VF Vulcan fiber

XLPE	Cross-linked polyethylene
XPS	Expandable or expanded polystyrene; (see also EPS)

INDEX

A

B

C

M

N

O

P